Comprehensive™

BIOTECHNOLOGY - III

Comprehensive™

BIOTECHNOLOGY - III

S. Mahesh

Asst. Professor, Deptt. of Biotechnology
R.V. College of Engineering
Bangalore

K.R.S. Murthy

Senior Lecturer, Deptt. of Biotechnology
Bangalore University, Bangalore

PUBLISHING FOR ONE WORLD

NEW AGE INTERNATIONAL (P) LIMITED, PUBLISHERS

(formerly Wiley Eastern Limited)

New Delhi • Bangalore • Chennai • Cochin • Guwahati • Hyderabad
Jalandhar • Kolkata • Lucknow • Mumbai • Ranchi

Visit us at www.newagepublishers.com

Published by New Age International (P) Ltd., Publishers
First Edition : 2006

ISBN : 81-224-1856-2

Rs. 90.00

C-06-05-735

Typeset at Pagitex Graphics, Delhi.
Printed in India at Mehra Offset, Delhi.

PUBLISHING FOR ONE WORLD

NEW AGE INTERNATIONAL (P) LIMITED, PUBLISHERS
(formerly Wiley Eastern Limited)
4835/24, Ansari Road, Daryaganj, New Delhi - 110002
Visit us at **www.newagepublishers.com**

Preface

Biotechnology is one of the fastest emerging fields in recent days due to its immense potential in various branches of Science and Technology. The benefit of Biotechnology to mankind is exemplary. It has acquired distinct academic and commercial interest in recent years. The potential implications of Biotechnology to other eminent branches like medicine, agriculture and environment has made it as one of the most attractive commercial ventures. Several fundamental subjects like Biochemistry, Molecular Biology and Biophysics have strengthened Biotechnology and contributed to its development to a greater extent. Understanding the basic functions and interactive role of biomolecules is indispensable, and plays a pivotal role in developing new techniques in Biotechnology.

The present book covers the syllabus of Third Semester Biotechnology—Biochemistry and Biophysics—prescribed by Bangalore University. The book is divided into two parts with thirteen chapters. Part A: Biochemistry covers seven chapters with suitable illustrations, and Part B: Biophysics covers six chapters with comprehensive examples.

This book endeavours to furnish simple, understandable text for third semester students of Biotechnology. One of the salient features of this book is that, Part A covers detailed Fundamentals of Biochemistry and Part B covers substantial information on biological examples at the backdrop of physics concepts.

We express our gratitude to Mr. Saumya Gupta, Managing Director, and V.R. Babu, Manager Marketing, New Age International Publishers, New Delhi, for their kind acceptance in publishing this book.

K.R.S. Murthy
S. Mahesh

Contents

Preface *v*

Part A: Biochemistry

1. Amino Acids **3–14**

1.1 Introduction 3
1.2 General Amino Acid Structure 3
1.3 Classification 4
1.3.1 Amino Acids with Non-Polar R-Groups 4
1.3.2 Amino Acids with Uncharged Polar R-Groups 5
1.3.3 Amino Acids with Positively Charged (Basic) R-Groups 5
1.3.4 Amino Acids with Negatively Charged (Acidic) R-Groups 6
1.4 Non-Standard Amino Acids 7
1.5 Stereochemistry of the Amino Acids 8
1.6 Acid-Base Properties 8

2. Proteins **15–24**

2.1 Introduction 15
2.2 Classification 15
2.2.1 Classification Based on Composition 15
2.2.2 Classification Based on Biological Function 16
2.2.3 Classification Based on Structure 17
2.3 Structural Organisation of Proteins 17
2.3.1 Primary Structure 18
2.3.2 Secondary Structure 19
2.3.3 Tertiary Structure 23
2.3.4 Quaternary Structure 23

3. Enzymes **25–41**

3.1 Introduction 25

3.2 Enzymes 26
3.2.1 Proteins 26
3.2.2 Biological Catalysts 26
3.2.3 Catalytic Efficiency 27
3.2.4 Specificity 27
3.2.5 Free Energy of Activation 28
3.2.6 Regulatory Control 29
3.2.7 The Active-Site 29
3.2.8 Binding Site 29
3.2.9 Catalytic Groups 29
3.3 Nomenclature 29
3.4 Classification 30
3.5 Chemical Nature and Properties of Enzymes 31
3.6 Monomeric and Oligomeric Enzymes 31
3.7 Factors Affecting Enzyme Activity 31
3.7.1 Effect of Enzyme Concentration 31
3.7.2 Effect of Substrate Concentration 31
3.7.3 Effect of Temperature 32
3.7.4 Effect of pH 33
3.7.5 Effect of Product Concentration 33
3.7.6 Effect of Activators 34
3.7.7 Effect of Time 34
3.8 Enzyme Kinetics Intramolecular Vs. Intermolecular Reactions 34
3.9 Derivation of Enzyme Kinetics Expressions 36
3.9.1 Rapid Equilibrium Assumption 36
3.9.2 Steady-State Approximation 37
3.10 Significance of Kinetic Constants 39
3.11 Lineweaver-Burk Double Reciprocal Plot 39
3.12 Coenzymes and Cofactors 40

4. Carbohydrates **42–62**

4.1 Introduction 42
4.2 Function 42
4.3 Classification 42
4.3.1 Monosaccharides 42
4.3.2 Oligosaccharides 43
4.3.3 Polysaccharides 43
4.4 Role of Carbohydrate in the Body 43
4.4.1 Energy Source 44
4.4.2 Protein Sparing 44
4.4.3 Metabolic Primer 44

4.4.4 Fuel for the Central Nervous System 44
4.5 Carbohydrates as a Source of Energy 44
4.5.1 Other Monosaccharides Enter the Glycolytic Pathway at Several Points 45
4.6 Stereochemistry of Monosaccharides 46
4.7 Fischer Projection Formula 48
4.8 Haworth Ring Structure 48
4.9 Conversion of Fischer Projection to Haworth Form 50
4.10 Conformation 51
4.11 Properties 52
4.12 Derivatives of Monosaccharides 56
4.13 Disaccharides and Oligosaccharides 58
4.14 Polysaccharides 60
4.14.1 Homopolysaccharides 61

5. Lipids **63–82**

5.1 Introduction 63
5.2 Properties 63
5.3 Lipids Functions 64
5.4 Classification 64
5.4.1 Fatty Acids 64
5.4.2 Lipids Containing Glycerol 68
5.4.3 Lipids Not Containing Glycerol 73
5.4.4 Lipids Combined with Other Classes of Compounds 81

6. Vitamins **83–94**

6.1 Classification of Vitamins 83
6.1.1 Fat-Soluble Vitamins 83
6.1.2 Water-Soluble Vitamins 89
6.1.3 Vitamin B Complex 89

7. Hormones **95–104**

7.1 Introduction 95
7.2 Classification 96
7.3 Steroid Hormone Biosynthesis 96
7.4 Steroids of the Adrenal Cortex 96
7.5 Gonadal Steroid Hormones 97
7.6 Function 98
7.6.1 Gonadal Steroid Hormones 98
7.6.2 Adrenocorticoid Hormones 99

Part B: Biophysics

1. **Scope and Development of Biophysics** **105–107**

1.1 Development of Biophysics 105
1.2 Scope 106
1.2.1 Understanding Molecular Spectrum 106
1.2.2 Microbiology 106
1.2.3 Molecular Biology 106
1.2.4 Medicine 106
1.2.5 Agriculture 107

2. **pH and Buffer Concepts** **108–115**

pH [Hydrogen-ion exponent] 109
Determination of pKa Values of Weak Acids 110
Buffers 112
How Does a Buffer Solution Work 114

3. **Chemical Bonding** **116–125**

3.1 Structure of Atom 116
3.2 Chemical Bond 116
3.2.1 Covalent Bond 117
3.2.2 Weak Bonds Stabilize the Structure of Biomolecules 118
3.2.2 Nature of Structural Bonds Between Proteins 120
3.3 Principles of Thermodynamics 123
3.3.1 Law of Thermodynamics 123
3.3.2 The Cell and Thermodynamics 125

4. **Analytical Techniques** **126–133**

4.1 Thin Layer Chromatography 126
4.1.1 Sample Application 126
4.1.2 Development of Chromatogram 126
4.1.3 Two Dimensional Separation 127
4.1.4 Detection 127
4.2 Gas Liquid Chromatography (GLC) 127
4.2.1 Principle 127
4.2.2 Sample Preparation and Application 128
4.2.3 Detector System 128
4.2.4 Applications of GLC 129
4.3 Centrifugation 130
4.3.1 Centrifugal Sedimentation 130
4.3.2 Preparation Centrifugation 130

4.3.3 Density Gradient Centrifugation 131
4.4 Ultracentrifugation 131
4.4.1 Preparation Ultracentrifugation 132
4.4.2 Analytical Ultracentrifugation 132
4.4.3 Applications of Ultracentrifuge 133

5. Spectroscopy **134–157**

5.1 Basic Principles 134
5.2 Beer-Lambert Laws 135
5.3 Colorimetry 136
5.3.1 Principle 136
5.3.2 Applications 137
5.4 Principles of Thermodynamics 137
5.4.1 Laws of Thermodynamics 138
5.4.2 The Cell and Thermodynamics 139
5.5 Visible Ultraviolet Spectroscopy 140
5.5.1 Principle 140
5.5.2 Description of Instrument 140
5.5.3 Overview of Operation 142
5.6 Fluorescence Spectroscopy Spectrofluorimetry 143
5.6.1 Principle 143
5.5.2 Instrumentation 143
5.6.3 Applications 144
5.7 Infrared Spectroscopy (IR) 144
5.7.1 Principle 145
5.7.2 Instrumentation 145
5.7.3 Applications 146
5.8 Raman Spectroscopy 146
5.8.1 Principle 146
5.8.2 Instrumentation 147
5.8.3 Applications 147
5.9 Atomic Absorption Spectroscopy 147
5.9.1 Principle 147
5.9.2 Instrumentation 148
5.8.3 Applications of AAS 149
5.10 X-ray Crystallography 149
5.10.1 Principle 149
5.10.2 Technique 150
5.11 Crystallographic Elucidation of Molecules 151
5.11.1 Small Molecules 151
5.11.2 Macromolecules 152
5.12 Nuclear Magnetic Resonance (NMR) 153

5.12.1 Principle 153
5.12.2 Instrumentation 155
5.12.3 Applications of NMR 156

6. Radioisotope Techniques **158–165**

6.1 Radiation 158
6.1.1 Sources of Ionising Radiation 159
6.1.2 Types of Isotopes 159
6.2 Radioactive Decay 159
6.2.1 α-rays 159
6.2.2 β-rays 159
6.2.3 X-rays and γ-rays 160
6.2.4 Rate of Radioactive Decay 160
6.2.5 Radioactive Units 161
6.3 Production of Radioisotopes 161
6.4 Measurement of Radioactivity 161
6.4.1 Geiger Mueller Counter 162
6.4.2 Scintillation Counter 162
6.5 Applications of Radioisotopes 163
6.5.1 Isotopes in Tracer Technique 163
6.5.2 Isotopes in Radioimmuno Assay (RIA) 164
6.5.3 Isotopes in Organ Functions and Imaging 164
6.5.4 Radiation and Isotopes in Agriculture 165
6.5.5 Isotopes in Food Industry 165
6.5.6 Isotopes in Cancer Therapy 165
6.5.7 Autoradiography and Probes 165

Part A
Biochemistry

1

Amino Acids

1.1 INTRODUCTION

Amino acids are the alphabet of protein structure and determine many of the important properties of proteins. Amino acids derive their name from the presence of an amino and carboxyl group bonded to an α-carbon atom. Glycine was the first amino acid to be isolated in protein hydrolysate by H. Braconnot in 1820, and threonine is the last of the 20 standard amino acids isolated from hydrolysates of fibrin by W.C. Rose in 1935. Besides the 20 standard amino acids known as normal or primary amino acids, found as building blocks of proteins, many additional biologically occurring amino acids serve other functions in the cells. All the 20 standard amino acids, except proline have a free carboxyl group and a free unsubstituted amino group on the α-carbon atom.

1.2 GENERAL AMINO ACID STRUCTURE

The amino acids differ from each other in the structure of their distinctive side chains, called the R-groups.

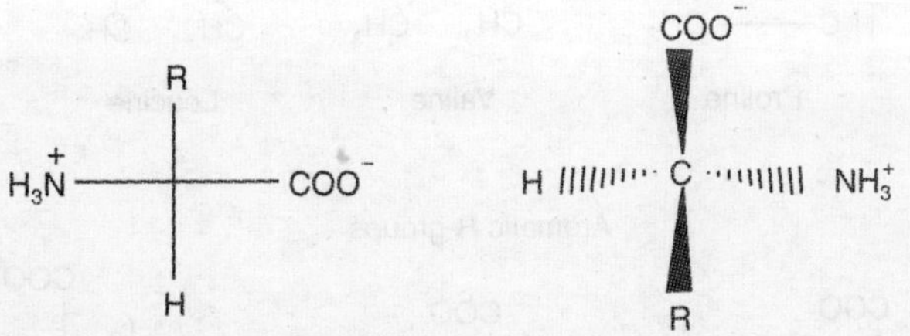

R
$H_3\overset{+}{N}$ — COO$^-$
H

COO$^-$
H ····· C ····· NH_3^+
R

Each of the 20 amino acids consists of two parts:

1. A part that is identical among all 20 amino acids; this part is used to link one amino acid to another to form the backbone of the protein.
2. A unique side chain (or "R-group") that determines the distinctive physical and chemical properties of the amino acid.

All amino acids consist of the following, bonded to a central or α-carbon:

CO_2H (carboxylic acid group or COO^- = carboxyl)

NH_2 group (amine group or NH_3^+ = amino)

Hydrogen atom

R-group or side chain differentiates the 20 amino acids

R-groups vary in size, charge, chemical reactivity, and Hydrogen-bonding capacity.

1.3 CLASSIFICATION

The amino acids are classified into four main classes based on the polarity of the R-groups:

1. Non-polar or hydrophobic R-groups
2. Neutral (uncharged) polar R-groups
3. Positively charged R-groups
4. Negatively charged R-groups (at pH 6 to 7, the zone of intracellular pH)

Within any single class of amino acids there are considerable variations in the size, shape and properties of the R-groups.

1.3.1 Amino Acids with Non-Polar R-Groups

Eight amino acids have non-polar R-groups. They include, five amino acids with aliphatic hydrocarbon R-groups (alanine, leucine, isoleucine, valine and proline), two with aromatic

Non-polar, aliphatic R-groups

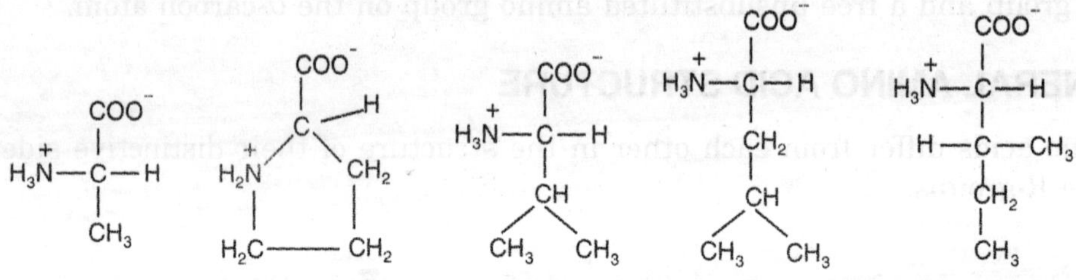

Alanine Proline Valine Leucine Isoleucine

Aromatic R-groups

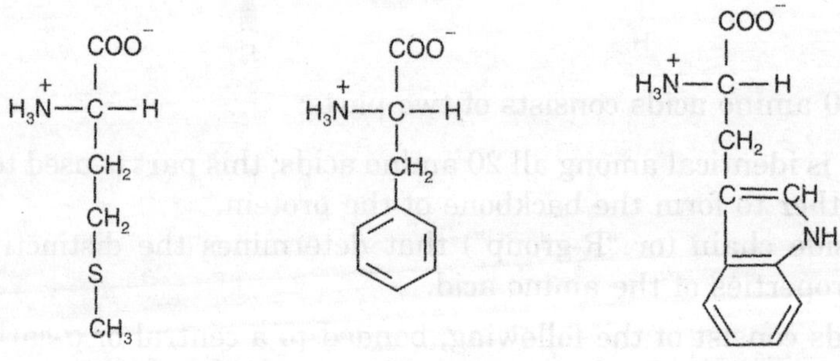

Methionine Phenylalanine Tryptophan

rings (phenylalanine and tryptophan), and one containing sulphur (methionine). The amino acids are less soluble in water than the amino acids with the polar R-groups. The least hydrophobic member of this class is alanine. Proline differs from all other amino acids in actually being an α-imino acid.

1.3.2 Amino Acids with Uncharged Polar R-Groups

These amino acids are relatively more soluble in water than those with non-polar R-groups. Their R-groups contain neutral (uncharged) polar functional groups which can hydrogen bond with water.

The polarity of serine, threonine, and tyrosine is contributed by their hydroxyl group, while that of asparagine and glutamine by their amide groups, and that of cysteine by its sulphydryl groups.

Glycine, the borderline member of this group, is sometimes classified as a non-polar amino acid, but its R-group, a single hydrogen atom, is too small to influence the high degree of polarity of the α-amino and α-carboxyl groups.

Polar, uncharged R-groups

Glycine Serine Threonine Cysteine Asparagine Glutamine Tyrosine

1.3.3 Amino Acids with Positively Charged (Basic) R-Groups

The basic amino acids have a net positive charge at pH 7 in their R-groups. All of them have six carbon atoms. They consist of lysine and arginine which bears a positively charged amino group at the ε-position on its aliphatic chain and positively charged guanidinium group, respectively.

Histidine contain weakly basic imidazolium function. Histidine is borderline in its properties. At pH 6, about 50% of histidine molecules possess a protonated, positively charged R-group, but at pH 7, less than 10% have a positive charge. It is the only amino acid whose R-group has a pK near 7.

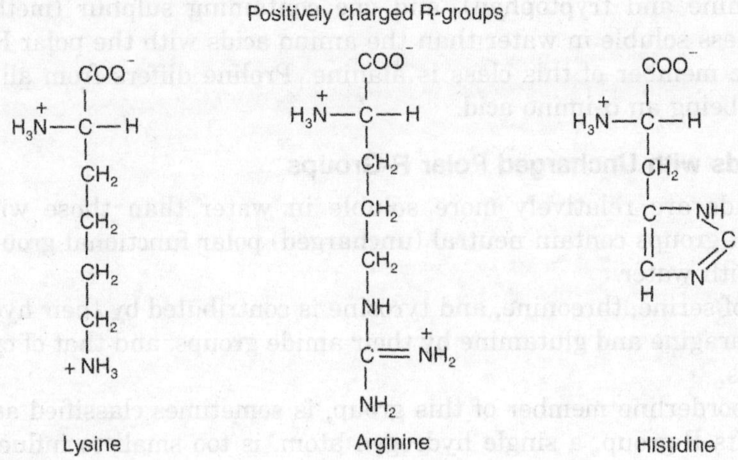

Positively charged R-groups

Lysine Arginine Histidine

1.3.4 Amino Acids with Negatively Charged (Acidic) R-Groups

The two members of this class are aspartic acid and glutamic acid, each with a second carboxyl group which is fully ionised and thus negatively charged at pH 6–7.

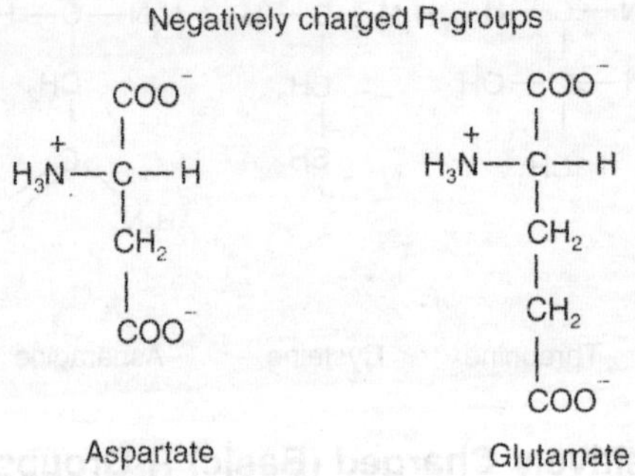

Negatively charged R-groups

Aspartate Glutamate

Amino acids are ordinarily designated by three letter symbols, but a set of one letter symbols has also been adopted to facilitate comparative display of amino acid sequences of homogeneous proteins.

AMINO ACID ABBREVIATIONS

Amino acid abbreviation	*3-letter abbreviation*	*1-letter abbreviation*	*Mnemonic for 1-letter*
Glycine	Gly	G	Glycine
Alanine	Ala	A	Alanine
Valine	Val	V	Valine
Leucine	Leu	L	Leucine

(Contd.)

Amino acid abbreviation	*3-letter abbreviation*	*1-letter abbreviation*	*Mnemonic for 1-letter*
Isoleucine	Ile	I	Isoleucine
Proline	Pro	P	Proline
Methionine	Met	M	Methionine
Phenylalanine	Phe	F	Fenylalanine
Tryptophan	Trp	W	tWyptophan (or tWo rings)
Tyrosine	Tyr	Y	tYrosine
Serine	Ser	S	Serine
Threonine	Thr	T	Threonine
Cysteine	Cys	C	Cysteine
Aspartic Acid	Asp**	D	asparDic acid
Glutamic Acid	Glu*	E	gluEtamic acid
Asparagine	Asn**	N	asparagiNe
Glutamine	Gln*	Q	Q-tamine
Histidine	His	H	Histidine
Lysine	Lys	K	(before L)
Arginine	Arg	R	aRginine

* Glx = either acid or amide (when it isn't known which it is)

** Asx = either acid or amide (when it isn't known which it is)

1.4 NON-STANDARD AMINO ACIDS

In addition to the 20 standard amino acids, proteins may contain non-standard amino acid residues, created by the modification of standard residues already incorporated into a polypeptide. All are derivatives of some standard amino acids.

$$\overset{+}{H_3N}-CH(CH_2OPO_3^{2-})-C(=O)-O^-$$

$$\text{4-hydroxyproline: } \overset{+}{H_2N} \text{ ring with } C(=O)-O^- \text{ and } OH$$

$$\overset{+}{H_3N}-CH(CH_2-CH_2-HC(OH)-CH_2-NH_2)-C(=O)-O^-$$

$$\overset{+}{H_2N}-CH(CH_2-CH(C(=O)O^-)_2)-C(=O)-OH$$

O-Phosphoserine　　4-Hydroxyproline　　5-Hydroxylysine　　carboxyglutamate

1.5 STEREOCHEMISTRY OF THE AMINO ACIDS

All amino acids contain α-carbon atom which is asymmetric (has four different substituents) except for one amino acid, for which the R-group is a hydrogen atom (glycine). The amino acids occur as enantiomers (non-superimposable complete mirror images) of each other. L-amino acids are the naturally occurring enantiomers found in all proteins. There are naturally occurring D-amino acids, but not in proteins (found in some bacterial cell wall peptide structures, in some peptide antibiotics, etc.).

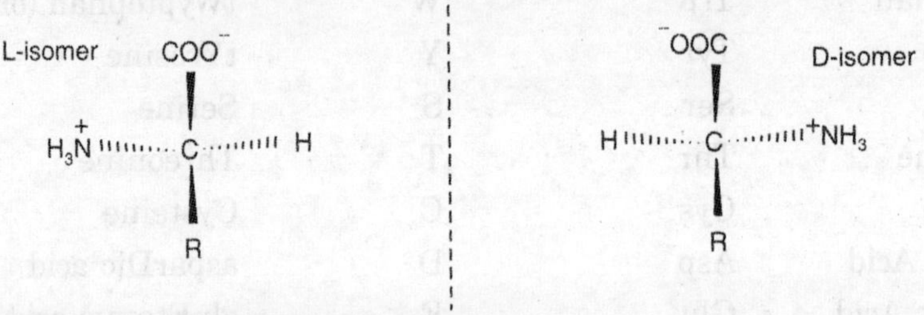

L-isomer COO^- — $H_3\overset{+}{N}$ ····· C ····· H — R

^-OOC D-isomer — H ····· C ····· $^+NH_3$ — R

Absolute configurations of D-glyceraldehydes is taken as the reference compound for α-amino acids. D- and L- apply only to the absolute configuration around the chiral α-carbon; 2 of the 20 amino acids (threonine and isoleucine) have a second chiral center, requiring the RS system to describe their structures accurately.

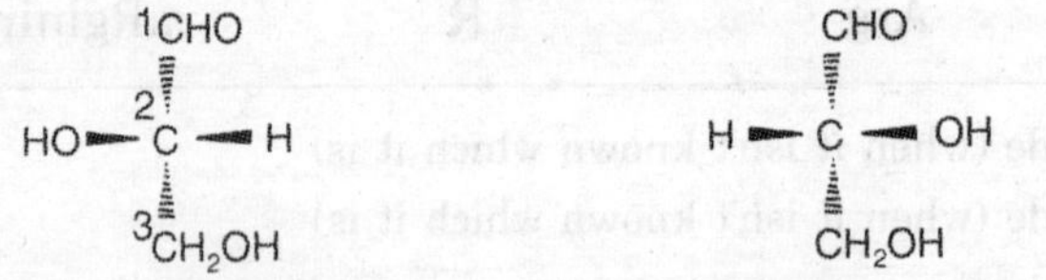

^{1}CHO — HO — ^{2}C — H — $^3CH_2OH$

L-Glyceraldehyde

CHO — H — C — OH — CH_2OH

D-Glyceraldehyde

COO^- — $H_3\overset{+}{N}$ — C — H — CH_3

L-Alanine

COO^- — H — C — $\overset{+}{N}H_3$ — CH_3

D-Alanine

1.6 ACID-BASE PROPERTIES

Amino acids are amphoteric and have characteristics of weak acids/bases (*i.e.*, can accept or donate protons).

$$R-COOH \longleftrightarrow R-COO^- + H^+$$

R – COOH acts as a H^+ donor in the reaction to the right

R – COO⁻ acts as a H⁺ acceptor in the reaction to the left

pK_a of a – carboxyl is 2.35 (at pH 7 = –1 charge)

$$R-NH_3^+ \longleftrightarrow R-NH_2 + H^+$$

R – NH_3^+ acts as a H^+ donor in the reaction to the right

R-NH_2 acts as a H^+ acceptor in the reaction to the left

pK_a of amino is 9.69 (at pH 7 = +1 charge)

Most of the amino acids are diprotic with two functional groups, but several AA's also have a 3rd functional group and are triprotic or *polyprotic* (only 7 of the 20 amino acids side chain contain ionizable groups). At low pH, both acid-base groups are fully protonated, so that it assumes the cationic form.

Amino acids in solution at neutral pH are thus predominantly dipolar ions or zwitterions. Zwitterions have a net charge of zero. The pH at which a molecule carries no net electric charge is known as its isoelectric point, pl. If pH < pl, net charge is positive (more + than – charges) and if pH > pl, net charge is negative (more – than + charges). pl is the pH exactly halfway between the two pK_a values surrounding the zero net charge equivalence point on the titration curve. The pH at which a compound exists as a zwitterion is its isoelectric point

fully protonated (+) ⟷ zwitterion (–/+) ⟷ fully deprotonated (–)

pk's of important groups (in solution)

Terminal carboxyl	2.3
Terminal amino	9.8
Carboxyl on asp and glu	4.0
Amine on lysine	10.5
Amine on arginine (Guanidino)	12.5
Hydroxyl on tyrosine	10.5
Hydroxyl on ser and thr	13.0
Sulphydryl on cysteine	8.5
Imidizole on histidine	6.0

Titration curve of glycine

$$\overset{+}{NH_3}-CH_2-COOH \underset{}{\overset{pK_1}{\rightleftharpoons}} \overset{+}{NH_3}-CH_2-COO^- \overset{pK_2}{\rightleftharpoons} NH_2-CH_2-COO^-$$

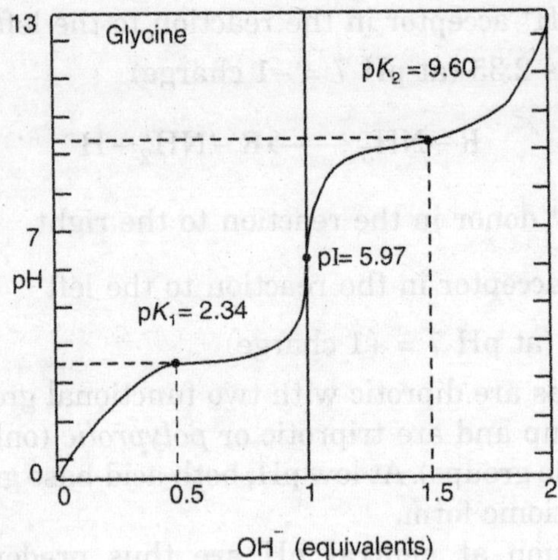

Titration curve of histidine

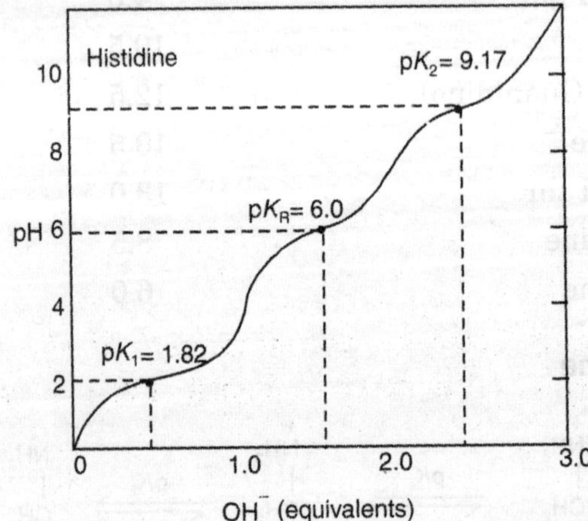

All amino acids have at least two reactive groups: α-amino and α-carboxyl groups and some amino acids have functional groups. These groups can react with a variety of reagents.

$$R{-}\underset{\underset{H}{|}}{\overset{\overset{NH_2}{|}}{C}}{-}COOH + NH_3 \longrightarrow R{-}\underset{\underset{H}{|}}{\overset{\overset{NH_2}{|}}{C}}{-}CONH_2 \quad \text{Amide formation}$$

$$R{-}\underset{\underset{COOH}{|}}{\overset{\overset{H}{|}}{C}}{-}NH_2 + RCHO \longrightarrow R{-}\underset{\underset{COOH}{|}}{\overset{\overset{H}{|}}{C}}{-}N{=}\underset{H}{C}{-}R \quad \text{Schiff base formation}$$

The Peptide Bond

Peptides and proteins are polymers of amino acids joined by peptide bonds. The condensation of α-carboxyl group of one amino acid with α-amino group of another amino acid in amide linkages results in the peptide bond. The process is repeated many times to form linear chain of amino acids — a polypeptide chain.

$$\overset{+}{H_3N}{-}\overset{\overset{R^1}{|}}{CH}{-}\underset{\underset{O}{\|}}{C}{-}OH + H{-}\overset{\overset{H}{|}}{N}{-}\overset{\overset{R^2}{|}}{CH}{-}COO^-$$

$$H_2O \rightleftharpoons H_2O$$

$$\overset{+}{H_3N}{-}\overset{\overset{R^1}{|}}{CH}{-}\underset{\underset{O}{\|}}{C}{-}\overset{\overset{H}{|}}{N}{-}\overset{\overset{R^2}{|}}{CH}{-}COO^-$$

Reactions of Amino Group

Ninhydrin Reaction: Ninhydrin (triketohydrindene hydrate) is an oxidating agent which leads to the oxidative deamination of alpha-amino groups to yield ammonia, carbondioxide and an aldehyde and the reduced form of ninhydrin—hydrindantin. Reduced ninhydrin reacts with an additional mole of ninhydrin in the presence of ammonia and gives a blue or purple colored product which is very specific and shows an absorption maximally at 570 nm. Imino acids (proline) yield a yellow product (absorption maximum 400 nm). The reaction is important for the detection and the quantitative analysis of amino acids.

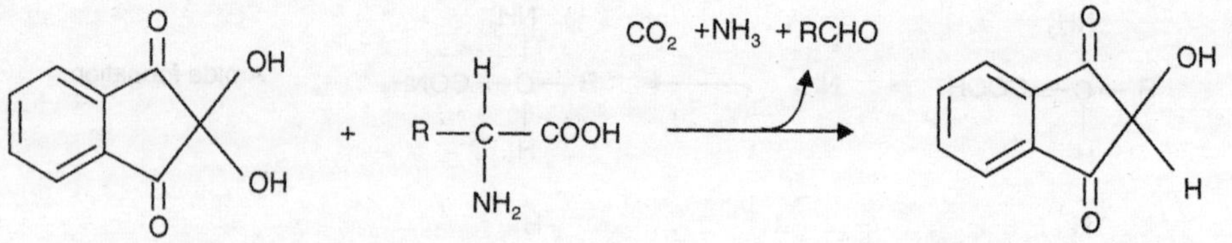

CO_2 $+NH_3$ + RCHO

Ninhydrin + Amino acid (R—C(H)(NH$_2$)—COOH) → Hydrindantin

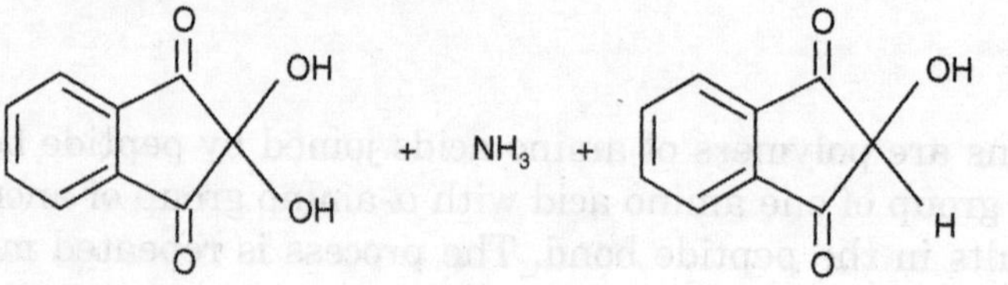

Ninhydrin + NH_3 + Hydrinadantin

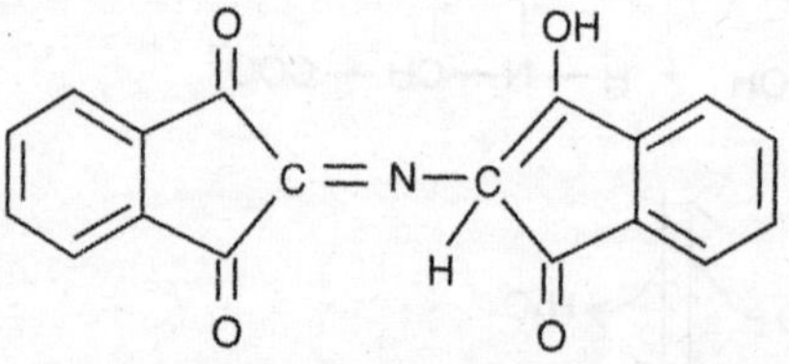

λ_{max} = 570 (Purple)

Sanger Reagent (FDNB): The arylation (dinitrophenylation, DNP) reaction of amino groups with FDNB (1-Fluoro-2, 4-dinitrobenzene) was used by Sanger *et al*., (1955), for the determination of the primary structure of insulin. The DNP-derivatives are stable to acid hydrolysis.

The reaction of FDNB with peptides/proteins leads to the formation of DNP-derivative of the terminal α - amino group and the DNP-derivative of other reactive groups (*e.g.*, the ε-amino group of lysine). After acid hydrolysis, only one amino acid modified at the N-terminal amino acid residue is produced. The DNP derivatives after acid hydrolysis can be extracted with organic solvents and identified by chromatography.

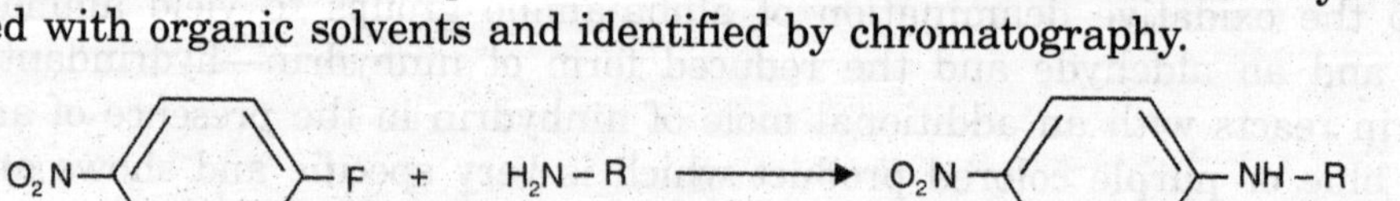

1-Fluoro-2, 4-Dinitrobenzene + H_2N-R → Yellow Product (O_2N—C$_6$H$_3$(NO$_2$)—NH−R)

Dansyl Chloride: Dansyl chloride (1-dimethylaminonapthalene-5-sulphonyl chloride) is useful for the determination of the N-terminal amino acid residue in peptides and proteins. The dansyl derivative is highly fluorescent (much more sensitive than FDNB) and stable to acid hydrolysis.

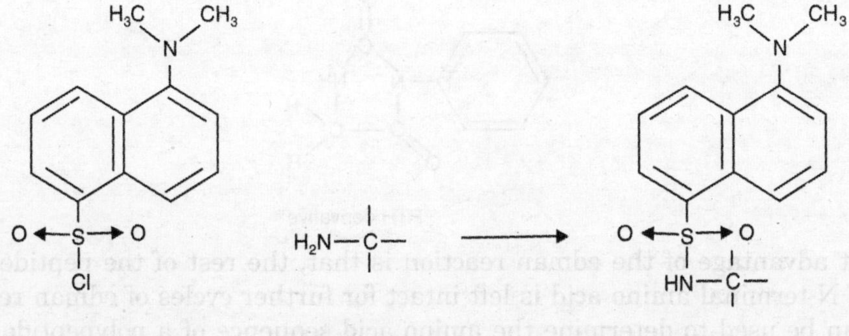

Dansyl Chloride Amino group

Edman Degradation: The most important and widely used labeling reaction for the determination of N-terminal residue is edman reaction. Phenylisothiocyanate (Edman Reagent) reacts quantitatively with the free amino group of a peptide to yield the corresponding phenylthiocarbamoyl derivative (PTC derivative).

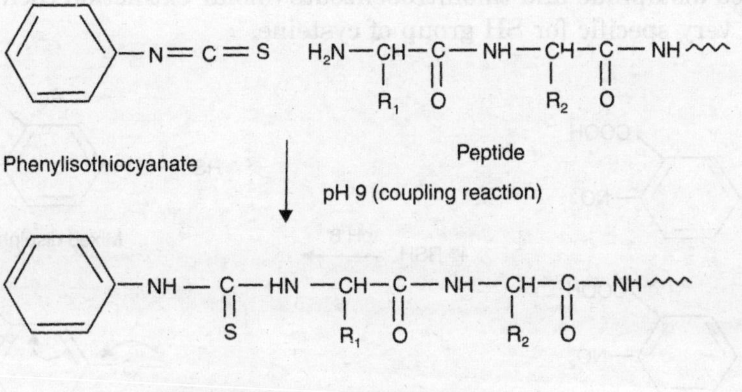

Phenylisothiocyanate Peptide

pH 9 (coupling reaction)

Phenylthiocarbamyl (PTC) derivative

After treatment with anhydrous trifluoroacetic acid (TFA, F_3CCOOH) the peptide bond involving the carboxyl of the N-terminal residue is cleaved (cleavage reaction) with the formation of the thiazolinone derivative. The treatment with TFA does not affect the other peptide bonds leaving a peptide chain with n-1 amino acid residues.

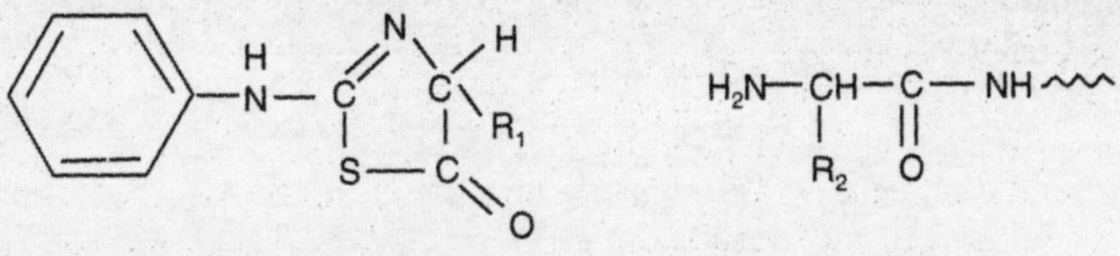

Thiazolinone derivative

The treatment of the thiazolinone derivative with aqueous acid (conversion reaction) leads to the formation of Phenylthiohydantoin (PTH) derivative of N-terminal amino acid which can be identified by chromatography.

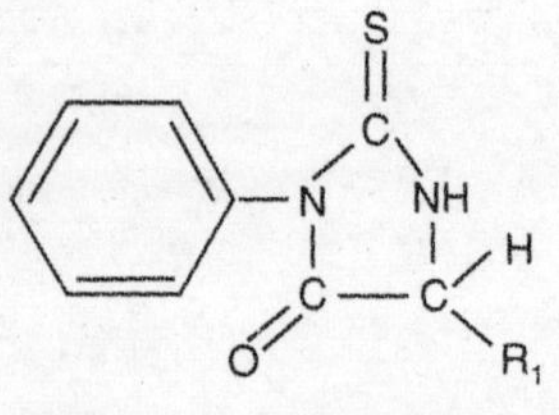

PTH derivative

A great advantage of the edman reaction is that, the rest of the peptide chain after removal of N-terminal amino acid is left intact for further cycles of edman reaction. This reaction can be used to determine the amino acid sequence of a polypeptide chain. This great advantage was further exploited by Edman & Begg and they perfected an automated amino acid sequenator for carrying out sequential degradation of peptides by the phenylisothiocyanate procedure.

Automated amino acid sequencers, now widely used, permit very rapid determination of the amino acid sequence of peptides upto 20 residues.

Elman Reagent: DTNB [5, 5′ - dithio - bis - (2-nitrobenzoate)] reacts with -SH groups to yield a mixed disulphide and thionitrobenzoate (molar extinction coefficient = 13600 at 412 nm). It is very specific for SH group of cysteine.

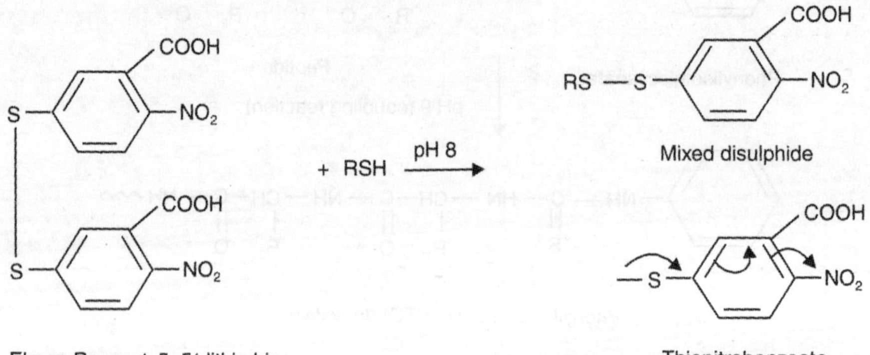

Elman Reagent: 5, 5′-dithio-bis-(2-nitrobenzoate) (DTNB) + RSH $\xrightarrow{\text{pH } 8}$ Mixed disulphide + Thionitrobenzoate

Elman Reagent: 5, 5′-dithio-bis-(2-nitrobenzoate) (DTNB)

Mixed disulphide

Thionitrobenzoate
ε = 13600 at 412 nm

2

Proteins

2.1 INTRODUCTION

The term protein is derived from the greek work proteios meaning "of prime importance" or "primary" or "holding first place" with respect to the cell constituents. Proteins are the most abundant organic molecules in the cells constituting 50% or more of their dry weight. They are found in every part of the cell, as they are fundamental in all aspects of cell structure and function. There are a huge variety of proteins, each specialised for a different biological function.

Amino acids are the building blocks of proteins. An amino acid consists of amino group, a carboxyl group, hydrogen atom and a functional R-group, which are bonded to an α-carbon atom. In proteins, the α-amino group of one amino acid is joined to the α-carboxyl group of another amino acid by a peptide bond. All proteins, whether obtained from the bacteria or from the most complex form of life, contain only a set of 20 amino acids, called standard amino acids.

2.2 CLASSIFICATION

Proteins are classified based on their composition, function, and conformation or structure.

2.2.1 Classification Based on Composition

Proteins are divided into two major classes on the basis of their composition – Simple proteins and Conjugated proteins.

Simple proteins are those which on hydrolysis yield only amino acids and no other major organic or inorganic hydrolysis products. They usually contain about 50% carbon, 7% hydrogen, 23% oxygen, 16% nitrogen and 0–3% sulfur.

Conjugated proteins are those which on hydrolysis yield not only amino acids but also organic or inorganic components. The non-amino acid part of a conjugated protein is called prosthetic group. Conjugated proteins are classified on the basis of the chemical nature of their prosthetic groups.

Class	*Prosthetic group*	*Example*
Lipoprotein	Lipid	β-lipoprotein of blood
Glycoprotein	Carbohydrate	Immunoglobulin G
Phosphoprotein	Phosphate group	Casein of milk
Hemoprotein	Heme (Iron Porphyrin)	Hemoglobin
Flavoprotein	Flavin nucleotides	Succinic dehydrogenase
Metalloprotein	Iron	Ferritin
	Zinc	Alcohol dehydrogenase
	Calcium	Calmodulin
	Copper	Plastocyanin
	Molybdenum	Dinitrogenase
Nucleoprotein	RNA	Ribosome

2.2.2 Classification Based on Biological Function

Proteins have many different biological functions, and are classified based on their function.

Types and Examples	*Function*
Enzymes	
Hexokinase	Phosphorylates glucose
Lactate dehydrogenase	Oxidises lactate
Storage proteins	
Ovalbumin	Egg-white protein
Casein	Milk protein
Gliadin	Seed protein of wheat
Zein	Seed protein of corn
Contractile proteins	
Myosin	Thick filaments in myofibril
Actin	Thin filaments in myofibril
Transport proteins	
Hemoglobin	Transports oxygen in blood of vertebrates
Myoglobin	Transports oxygen in muscle cells
Serum albumin	Transports fatty acids in blood
Lipoprotein	Transports lipids in blood
Hormones	
Insulin	Regulates glucose metabolism
Adrenocorticotropic hormone	Regulates corticosteroid synthesis

(Contd.)

Types and Examples	*Function*
Protective proteins in vertebrate blood	
Antibodies	Form complexes with foreign proteins
Fibrinogen	Precursor of fibrin in blood clotting
Thrombin	Component of clotting mechanism
Toxins	
Ricin	Toxic protein of castor bean
Diphtheria toxin	Bacterial toxin
Clostridium botulinum toxin	Causes bacterial food poisoning

2.2.3 Classification Based on Structure

In the native state, each type of protein molecule has a characteristic three-dimensional structure, referred to as its conformation. Depending on their conformation, proteins are classified as Fibrous and Globular Proteins.

Fibrous proteins consist of polypeptide chains arranged in parallel along a single axis to yield long fibers or sheets. Fibrous proteins are insoluble in water or dilute salt solutions. They are the structural elements in the connective tissue of higher animals. For example, collagen of tendons and bone matrix, elastin of elastic connective tissue, α-keratin of hair, horn, skin, nails, feathers, etc.

Globular proteins consist of polypeptide chains tightly folded into compact spherical or globular shapes. Most globular proteins are soluble in aqueous solutions. They have a mobile or dynamic function in the cell. Of the nearly 2000 different enzymes known to-date, nearly all are globular proteins.

Some proteins fall between the fibrous and globular types, resembling fibrous proteins in their long rod-like structures and the globular proteins in their solubility in aqueous salt solutions. For example, myosin, an important structural element of muscle and fibrinogen, the precursor of fibrin, the structural element of blood clots.

2.3 STRUCTURAL ORGANISATION OF PROTEINS

The function of protein can be understood only in terms of three dimensional structure of proteins. The structural descriptions of proteins are described in terms of four level of organisations:

1. Primary structure — the amino acid sequence in the polypeptide chain.
2. Secondary structure — the local spatial arrangement of a polypeptide backbone atoms, without regard to the arrangement of amino acids side chain. This refers to the α-helices or β-pleated sheet, and the random coil structure.
3. Tertiary structure — the three dimensional structure of an entire polypeptide chain (polypeptide backbone and amino acid side chain).
4. Quaternary structure — the spatial arrangement of subunits. Many proteins are composed of two or more polypeptide chains, loosely referred to as subunits, which associate through non-covalent interactions, and in some cases covalently associated through disulphide bonds.

2.3.1 Primary Structure

Amino acids are the building blocks of proteins. The amino acids are held together in a protein by covalent bond which are known as peptide bonds or linkages. A peptide bond is formed by the condensation of the amino group of an amino acid with the carboxyl group of another amino acid. A dipeptide will have two amino acids, but contains one peptide bond. Peptides containing more than 10 amino acids are referred to as polypeptides.

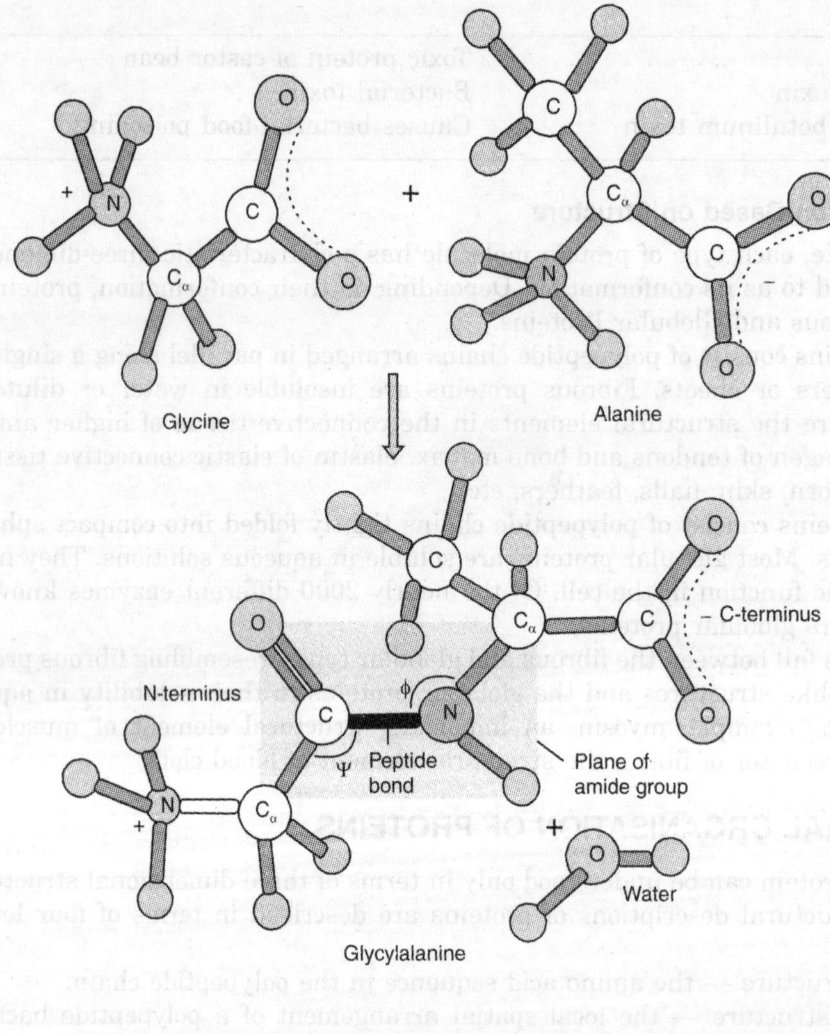

Glycine

Alanine

Glycylalanine

N-terminus

C-terminus

Peptide bond

Plane of amide group

Water

The peptide bond has partial double bond character

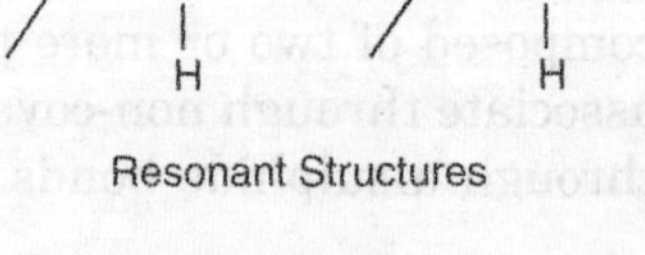

Resonant Structures

Linus pauling and Robert Corey, in late 1930's analysed the peptide bond. The α-carbon of adjacent amino acid residues are separated by 3 covalent bonds, arranged as – Cα – C – N – Cα – C – N -. The peptide bond in shorter than the C – N bond in a simple amine. The atoms associated with peptide are coplanar indicating a resonance structure. The six atoms of the peptide lie in a single plane, with the oxygen atom of the carbonyl group and the hydrogen atom of the amide nitrogen trans to each other. The peptide bonds due to partial double bond character are unable to rotate freely. Rotation is permitted about N – Cα (φ) and the Cα – C bonds (ψ). Both the – C = O (carbonyl group) and – NH groups of peptide bonds are polar and are involved in hydrogen bond formation.

The peptide chains are written with the free amino end (N – terminal residue) at the left, and the free carboxyl end (C – terminal residue) at the right. The amino acid sequence is read from the N – terminal end to the C – terminal end. The amino acids in a peptide or protein are represented by the three-letter or one-letter abbreviation.

The amino acid sequence in the polypeptide chain represents the primary structure. The bovine polypeptide hormone, insulin, was the first protein for which complete amino acid sequence was determined by Frederick Sanger in 1953. The elucidation of the primary structure of 51 amino acid residue, insulin, was the labor of many scientists over the period of a decade, and they utilised ~ 100 gm protein.

Insulin consists of two chains, linked by disulfide bridges. Chain A has 21 amino acid residues while chain B has 30 amino acid residues. The two chains are bound together in their quaternary structure by two disulphide bridges. In addition, a third disulphide bridge between two amino acids in the "A" chain help stabilise its tertiary structure.

30
Ala.Lys.Pro.Thr.Tyr.Phe.Phe.Gly
Arg
Glu
NH_2 NH_2
Chain B 1 Phe.Val.Asp.Glu.His.Leu.Cys.ly.Ser.His.Leu.Val.Glu.Ala.Leu.Tyr.Leu.Val.Cys.Gly
S S
NH_2 S NH_2 NH_2 S NH_2
Chain A 1 Gly.Ile.Val.Glu.Glu.Cys.Cys.Ala.Ser.Val.Cys.Ser.Leu.Tyr.Glu.Leu.Glu.Asp.Tyr.Cys.Asp 21
S — S

2.3.2 Secondary Structure

The secondary structure refers to the local conformation of its backbone. It is the precise and repeating folding due to hydrogen bonding with respect to the amino acid backbone into a helix or a β-pleated sheet and turns and random coil structure. The most common structure is the α-helix and β-pleated sheet. This structure is a pleated sheet formed by parallel chains of amino acids. These sheets are important in many structural proteins. Many proteins have sheets and helices. Secondary structure arises from the geometry of the bond angle between amino acids as well as hydrogen bonds between nearby amino acids.

2.3.2.1 Helical Structure

The polypeptide chain is twisted by the same amount about each of its C_α atoms and the chain assumes a helical conformation. A helix is characterised by the number, *n*, of peptide units per helical turn, and its pitch, *p* — the distance, the helix rises along its axis per turn. The helix has chirality, and it may be either right-handed or left-handed. The helical structure is stabilised by hydrogen bonds. Pauling and Corey in 1951 discovered the α-helix (Nobel Prize, 1954) through model building. The polypeptide backbone is tightly wound around an imaginary axis drawn longitudinally through the middle of the helix and the R-groups of the amino acid residues protrude outward from the helical backbone. They studied the flexibility in the covalent bonds of a peptide backbone structure as well as the partial charges of the backbone structure, and determined that peptides could bend EVERY FOURTH (actually every 3.4) amino acid (alpha helix). Alpha means it twists in a right handed (clockwise manner). The α-helix is the predominant structure in α-keratins, such as hair.

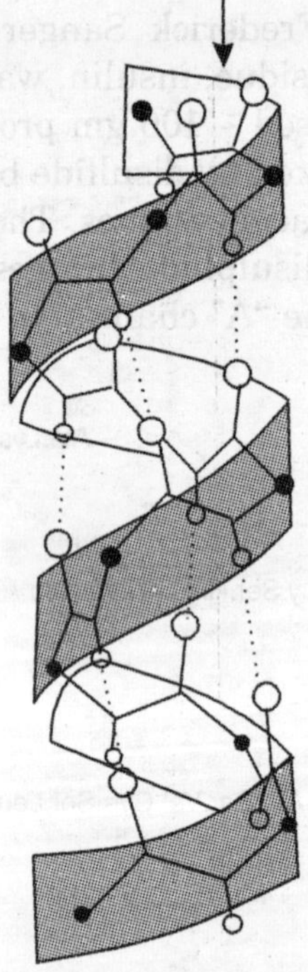

Secondary structure (helix)

A typical growth rate for human hair is about one half of an inch per month. This requires that the hair follicle produce approximately ten turns of alpha helical protein every second. Ten turns, by the way, is about thirty-six amino acids. The protein strands within hair and other alpha-keratins are crosslinked to some extent by covalent bonds between cysteine residues to form disulphide bonds. The more such disulphide bonds there are between the strands, the more rigid the protein becomes as a whole. The alpha keratins can be classed as "soft" or "hard" according to their sulphur content, which is to say, the relative number of cysteines in the polypeptide chains. The low-sulphur keratins of skin and callous are much more flexible than the high-sulphur, hard keratins of horns, claws and hooves.

Cysteine-Cystine Transformation

The basic principle of the permanent wave process for hair involves breaking the existing disulphide bonds between α-helices and then reforming new disulphide bonds after the hair fibers have been shaped and rearranged by the hair stylist. A reducing agent is first applied to the hair to break the disulphide bridges. The hair is then arranged into the desired shape and an oxidising agent applied to cause reformation of the disulphide bonds. A "permanent" is really only permanent for the portion of the hair that was processed, and it lasts until new, untreated keratin replaces it.

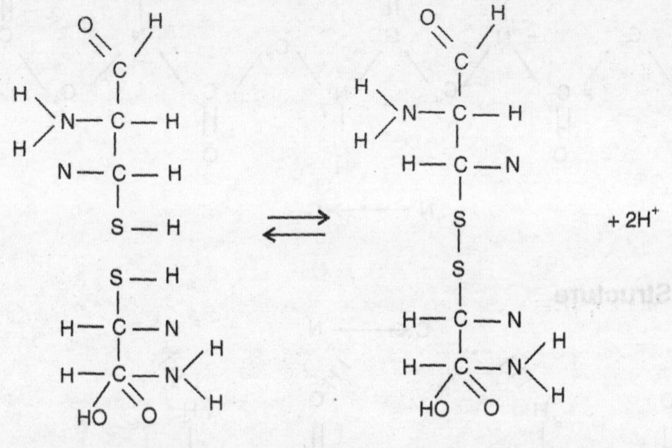

2 Cysteine $\longrightarrow$ Cystine $+ 2H^+$

2.3.2.2 β-Pleated Sheet

Linus Pauling and Robert Corey postulated the existence of a second type of secondary structure, the β-pleated sheet. This conformation utilises the full hydrogen bonding capacity of the polypeptide backbone, but the hydrogen bonding occurs between the neighbouring polypeptide chains rather than within the same polypeptide chain as in the case of α-keratin. β-pleated sheets are of two types:

1. The antiparallel β-pleated sheet, in which neighbouring hydrogen bonded polypeptide chains run in opposite directions
2. The parallel β-pleated sheet, in which neighbouring hydrogen bonded polypeptide chains extend in the same direction

β-pleated sheets are common structural motifs in proteins. In globular proteins, they consist of 2 to as many as 15 polypeptide strands, the average being 6 strands. Each polypeptide chain in a β-sheet contains upto 15 amino acid residues, with the average being 6 amino acid residues. Parallel β-sheets of less than 5 strands are rare and are less stable than antiparallel β-sheets. This is due to the distortion of hydrogen bonds in parallel β-sheets in comparison to those of antiparallel β-sheets. Mixed parallel- antiparallel β-sheets are commonly found in proteins.

Antiparallel β-pleated Structure

Parallel β-pleated Structure

2.3.2.3 Random Coil Structure

Regular secondary structures — helices and β sheets comprise around half of the average globular protein. The remaining polypeptide segment has a loop or coil conformation. These structures are orderly structures just like helices or sheets. In contrast, random structure or coil structure refers to the totally disordered and rapidly fluctuating set of conformations assumed by denatured proteins in solution.

Many proteins have regions that are truly disordered, and often wave around in solution because there are few forces to hold them in place. Sometimes, entire polypeptide chain segments are disordered. These may play a role in the binding of specific molecules.

2.3.3 Tertiary Structure

Tertiary structure refers to a higher level of folding in which the helices and sheets of the secondary structure fold upon themselves. This higher level folding arises for several reasons. First, different regions of the amino acid chain are hydrophilic or hydrophobic and arrange themselves accordingly in water. Second, different regions of the chain bond with each other via hydrogen bonding or disulfide linkages.

Myoglobin is the first protein whose tertiary structure was established by Kendrew (Nobel Prize 1962) using X-ray diffraction. Most water soluble-proteins have a hydrophobic interior and a hydrophilic exterior.

A common way for enzymes to denature is to unfold either because of hydrogen bond breakage (often due to pH or temperature), or oxidants or reductants that unnaturally break or form disulphide bridges. In any case, the active site is affected, which in turn affects the activity. The final protein structure depends upon how the protein folds as it comes off of the ribosome and any subsequent processing (*i.e.*, the functional folded protein may not be the way the protein would fold if it were completely unfolded and allowed to refold). Sometimes, there are chaperones that help, proteins fold correctly (after being made off of a ribosome or where, there is a tendency for denaturing). Chaperones themselves are proteins.

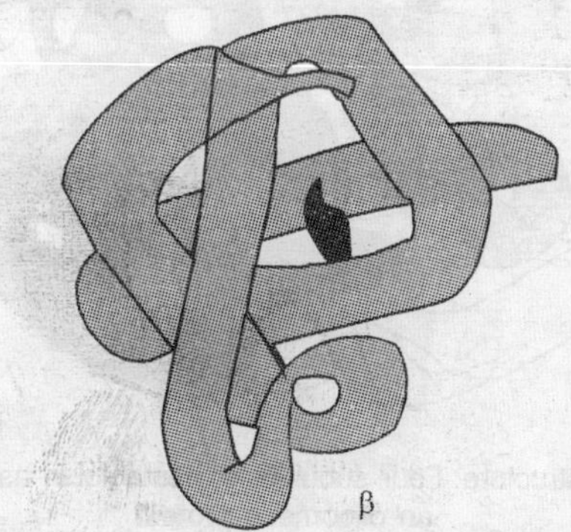

Tertiary structure: One complete protein chain (βchain of hemoglobin)

2.3.4 Quaternary Structure

Quaternary structure arises when different polypeptide chains or subunits are bound together usually by hydrogen bonds. For example, hemoglobin — the oxygen carrying protein in blood has four subunits hydrogen bonded together. Usually the function of the total quarternary structure is 'better' than the function of the sum of the individual protein chains. Most proteins with a molecular weight of 50,000 or more are made of such units.

Sometimes, quaternary structure maybe very complex. For example, beef glutamate dehydrogenase is an enzyme with a molecular weight of 2,200,000. Each enzyme molecule consists of eight large subunits. In turn, each of these consists of numerous smaller units. These polypeptide chains self-assemble into a complete and functional protein. The cell takes full advantage of this property to rapidly generate the cytoskeleton much of which consists of very long chains or helices, or tubes of proteins subunits.

Hemoglobin carries oxygen from the lungs to the tissue. Myoglobin performs a similar function in muscle tissue, taking oxygen from the hemoglobin in the blood and storing it or carrying it around until needed by the muscle cells. Hemoglobin and myoglobin also have similar structures. Myoglobin contains 151 amino acid residues plus a heme group to bond to oxygen. Hemoglobin has four chains, two with 141 residues and a heme group and two with 146 residues and a heme group. The molecular weight of hemoglobin is about 64,500 and can carry four oxygen molecules.

It is important that hemoglobin can bond to oxygen under certain conditions. But it is equally important that hemoglobin can release oxygen under other conditions. The ability of hemoglobin to bind oxygen is sensitive to several factors. They include pH, temperature, concentrations of O_2 and CO_2, and even the number of oxygen molecules already bound. When oxygen binds to hemoglobin, the structure of the hemoglobin changes slightly so that it binds more efficiently to oxygen, thus enhancing its ability to carry more oxygen.

Quaternary structure: Four subunits of hemoglobin assembled into an oligomeric protein

3

Enzymes

3.1 INTRODUCTION

Enzymes are biocatalysts synthesized by the living cells. They increase the rate or velocity of a reaction without itself undergoing any change in a biological system. Each reaction taking place in a cell is catalyzed by its own particular enzyme, so that in a given cell there are a large number of enzymes. In the absence of these enzymes, many of these reactions would not occur ever over a period of years, and life, as we know today would not exist.

Berzelius in 1836 coined the term catalysis (Greek:to dissolve) to substances that increase the rate of a chemical reaction, without themselves undergoing any change. In 1878, Kuhne used the word enzyme (Greek:in yeast) to indicate that catalysis takes place in biological systems. The term substrate refers to the reactants and was introduced by Duclaux in 1883.

There are many reasons for studying enzymes–their biological and chemical roles, medical, therapeutic and industrial uses, and their use as tools in molecular biology (rDNA and genetic engineering would not be possible without restriction endonuclease enzymes), to mention a few. Enzymes have been called the "agents of life", since all life processes are dependent on them.

The first clear realization of the existence of enzymes came in 1879 by E. Buchner. Until that time it was believed that only living cells could be biochemically functional. Buchner prepared an extract of yeast cells by grinding and filtering off the cell debris. To preserve the resulting clear, cell-free solution for future study he added sugar as a preservative (just as many food producers do today, to preserve food) and observed that fermentation occurred (formation of alcohol from sugar), indicating that living cells were not required for carrying out metabolic processes. Instead, some smaller entities must be present which could bring about the conversion of sugar to alcohol. These were named enzymes. A major conceptual breakthrough came a couple of decades later when Emil Fisher proposed the "lock and key" model, to explain the specificity of enzymes to a particular substrate. The substrate is analogous to the key and fits into the enzyme the same way a key fits into a lock.

3.2 ENZYMES

- Proteins (recent developments indicate that both RNA and antibodies may have catalytic activity, and these are called ribozymes, and catalytic antibodies or abzymes, respectively)
- Biological catalysts, critical components of cell metabolism & biological processes.
- Very efficient and specific catalysts
- Reduce ΔG‡ for reaction (by binding the transition state)
- Subject to regulatory control by various means
- Carry out catalysis in a special region of the molecule, called the active-site
- Exhibit special kinetics

3.2.1 Proteins

- All enzymes are proteins, with the exception of some small catalytic RNAs
- MW's range from 10^4 to 10^6 daltons
- May be single polypeptide chains, or oligomers of several subunits (most commonly oligomers are dimers, or tetramers. Some multienzyme complexes are known with as many as 48 subunits).
- May have more than one enzyme activity associated with the same protein (*i.e.*, there are some large enzymes which catalyze more than one reaction (frequently successive steps in a metabolic pathway)
- Often contain a prosthetic group (or cofactor): Typical examples are: metal ions, heme, Fe – S clusters, coenzymes (*e.g.*, NADH, FAD, FMN, PLP)
- Coenzymes usually are vitamins, or derived from vitamins, and act as carriers (*e.g.*, of H, e^-, CO_2).
- Enzymes are usually named after their substrate by adding *ase*, *e.g.*, protease (proteinase), esterase, α-glucosidase, alcohol dehydrogenase, β-lactamase.

3.2.2 Biological Catalysts

Catalysts speed the rate of attainment of equilibrium by lowering the energy barrier between substrate and products. In other words a catalyst will increase the rate of a reaction but not affect the position of the reaction equilibrium. The catalyst is not used up in the reaction but is regenerated.

Consider the reaction

$$\underset{\text{substrate}}{A} \underset{k_{-f}}{\overset{k_f}{\rightleftharpoons}} \underset{\text{product}}{B}$$

suppose that for the uncatalyzed reaction

$k_f = 10^{-4}\ s^{-1}$

and $k_{-f} = 10^{-8}\ s^{-1}$

then $K_{eq} = [B] / [A] = k_f / k_{-f} = 10^4$

i.e., the equilibrium favors the formation of B strongly. Starting with A, the reaction will take about 10 hr. to reach equilibrium.

k_{obs}, the observed rate constant for such a reaction is given by:

$$k_{obs} = k_f + k_{-f} = 10^{-4}\ s^{-1}$$

For a first order reaction the rate constant is related to the starting material by:

$$[A] = [Ao]\ e^{-kt}$$

$$\ln [Ao]/[A] = kt$$

where $A = Ao/2$, when $t = t_{1/2}$

Thus, $\ln 2 = kt_{1/2}$

$$t_{1/2} = \ln 2/k = 0.69/k$$

The reaction appears over after about five half-lives, *i.e.*, 3.5/k.

Consider now the catalyzed reaction: typical rate enhancements by enzymes over the corresponding uncatalyzed reaction are 10^6 to 10^{16}.

Thus, K is still 10^4 but both k_f and k_{-f} will be increased, let's say by about 10^8.

Hence, $k_f = 10^4\ s^{-1}$, and $k_{-f} = 1\ s^{-1}$

Thus, $K_{eq} = [B]/[A] = k_f/k_{-f} = 10^4$

k_{obs}, the observed rate constant for such a reaction is given by:

$$k_{obs} = k_f + k_{-f} = 10^4\ s^{-1}$$

Therefore, it will now only take about 0.35 msec to reach equilibrium.

Essentially, all life processes are controlled by enzymes. A typical cell will have several thousand different enzymes, carrying out a variety of different types of reactions. It is currently estimated that there are about 75,000–100,000 genes in the human genome—a substantial fraction corresponding to enzymes.

3.2.3 Catalytic Efficiency

Enzymes are unique in the great rate enhancements they bring about, orders of magnitude greater than other catalysts *e.g.*, Pt/C used in catalytic converters in cars and in the hydrogenation of fats and oils. Typical rate enhancements by enzymes over the corresponding uncatalyzed reaction are 10^6 to 10^{16}. A typical enzyme will convert around a thousand molecules of substrate to product in 1 second! Some will convert as many as a million!! There are of course slower enzymes as well.

Unlike most other catalysts, enzymes work at ambient temperature, atmospheric pressure, and usually in a narrow pH-range near neutrality (there are exceptions, *e.g.*, pepsin, in the stomach, which is a protease and breaks down proteins into peptides, operates in the pH 2–5 region).

3.2.4 Specificity

This is the second unique feature of enzymes as catalysts. They are very specific. A given enzyme will catalyze only one type of reaction for one type of reactant or substrate. They are also stereospecific, and produce no by-products *e.g.*, consider thrombin, an enzyme in the blood-clotting cascade, and glucose oxidase:

	Thrombin	*Glucose Oxidase*
Type of reaction catalyzed	hydrolysis	oxidation
Type of bond	peptide	sugar hemiacetal
Structure about bond	between Arginine and Glycine	glucose
Stereospecificity	both must be L-amino acids	D-configuration

Cascades are "nature's way" of amplifying signals. Blood-clotting must be very carefully controlled to prevent undesirable hemorrhaging or clotting. A cascade consists of a series of enzymes, each of which can exist in an active and inactive form. The amplification works by starting with them all in the inactive form. On triggering the system, each one in turn activates the next; thus very rapidly one gets tremendous amplification of the initial signal. Thrombin is the last enzyme in the clotting cascade. It is activated from prothrombin, and in turn converts fibrinogen, which is soluble, into insoluble fibrin (clots).

Due to the asymmetric active sites (a consequence of their three-dimensional nature), enzymes exhibit absolute stereospecificity; this was demonstrated in a classic experiment by Westheimer and Vennesland using yeast alcohol dehydrogenase (the corresponding liver enzyme tends to be overworked at holidays such as New Years—its the major enzyme involved in the metabolism of ethanol). They showed that the enzyme can distinguish between the two prochiral H's of ethanol:

$H_3C — C(H_a)(H_b) — OH$ $\quad$ $H_3C — C(H_a)(H_b) — OH$

The enzyme will stereoselectively remove H_a in the oxidation direction, and return it in the reverse reaction. (The reaction catalyzed is the oxidation of alcohol to acetaldehyde, or its reverse.)

3.2.5 Free Energy of Activation

Enzymes act as catalysts since they lower the free energy of activation (ΔG‡) for the reaction. They do this by a combination of raising the ground state ΔG of the substrate and lowering the ΔG of the transition state (TS) for the reaction, thereby, decreasing the barrier for reaction to occur. The presence of the enzyme leads to a new (different) reaction pathway than for the uncatalyzed reaction. The major way by which enzymes bring about their great rate enhancements is by tight binding of the TS.

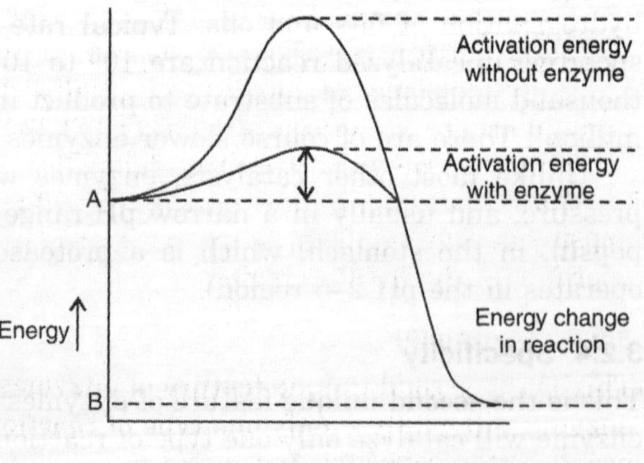

Fig. 3.1

The height of the energy barrier is the free energy of activation.

3.2.6 Regulatory Control

Enzymes control the flux of metabolites through metabolic pathways (*i.e.*, the amount and rate of material passing through the pathway). There are many levels at which such control is exerted; *e.g.*,

Enzyme concentration—at the genetic level

Feedback inhibition—the end product of a metabolic pathway inhibits (or activates) the enzyme, thereby decreasing (or increasing) the rate of formation of product by causing the enzyme to be less or more active than normal.

Covalent modification—many enzymes are activated or inactivated, *e.g.*, by phosphorylation.

3.2.7 The Active-Site

That region of the enzyme molecule where the substrate binds and the catalytic reaction occurs is the active site or active center.

- Usually a cleft or pocket on the surface of the enzyme, often at the interface of two domains.
- Usually involves only a small fraction of the enzyme surface.
- Is complementary to the substrate shape and polarity.
- Contains binding site for the substrate (attracts and positions substrate)
- Contains catalytic groups — these are the reactive side chains of amino acids or cofactors, which carry out the bond breaking / forming reactions.
- Binding involves non — covalent interactions such as H — bonding, electrostatic interactions, hydrophobic interactions, Van der Waals interactions.
- Rest of the protein structure provides – i) a superstructure to position the substrate and catalytic groups, ii) flexibility for conformational changes — a means for regulatory control and sites for recognition by other biomolecules.

3.2.8 Binding-Site

The shape and polarity accounts for much of the specificity of enzymes. There is complementarity between the shape and polarity of substrate and the active site. In some cases, binding of the substrate induces a conformational change. This is particularly common where there are two substrates. Binding of the first induces a conformational change which results in the formation of the binding site for the second substrate. Hexokinase is a good example of this (ATP + glucose ⟶ glucose phosphate + ADP).

3.2.9 Catalytic Groups

Usually the amino acid side chains and/or cofactors can act as catalysts:

acids: – COOH, – Imidazolium
bases: – NH2, imidazole, – S – (from Cys)
nucleophiles: imidazole, – S –, – OH
electrophiles: often metal ions *e.g.*, Zn^{+2}

3.3 NOMENCLATURE

As in the case of nomenclature of organic compounds during development of organic chemistry, many enzymes were given trivial names before any attempt was made at a systematic naming of enzymes. In general, the trivial names consisted of the suffix "– ase" added to the substrate acted on, as in the case of Urease – which catalyzes the hydrolysis of urea, or indicated about the reaction catalyzed, as in the case of Lactate dehydrogenase, which catalyzes the dehydrogenation of lactate to pyruvate. The only major exceptions to this are the proteolytic enzymes, whose names usually end with "in". For example, Trypsin, Papain, Chymotrypsin, etc.

As the list of known enzymes rapidly grew, there was a lack of consistency in the nomenclature. Hence, there was a need for a systematic way of naming and classifying the enzymes. The present day accepted nomenclature and classification is that recommended by the Enzyme Commission.

3.4 CLASSIFICATION

The Enzyme Commission divided enzymes into six main classes, on the basis of the reaction catalyzed. Each enzyme was assigned a code number (Enzyme Commission Number), consisting of four elements, separated by dots.

The first number indicates the type of reaction being catalyzed or the main class to which the enzyme belongs and it has the values from 1 to 6.

The second number indicates the subclass, which usually specifies the type of substrates or the bond cleaved.

The third number indicates the sub-subclass, and defines the reaction catalyzed.

The fourth number indicates the serial number of the enzyme in its sub-subclass.

First digit	*Enzyme class*	*Type of reaction catalyzed*
1	Oxidoreductase	Oxidation-Reduction reactions
2	Transferase	Transfer of an atom or group between two molecules
3	Hydrolase	Hydrolytic reactions
4	Lyase	Elimination reactions
5	Isomerase	Isomerization reactions
6	Ligase	Synthetic reactions in which two molecules are joined at the expense of an high energy molecule, such as ATP

The Enzyme Commission assigned to each enzyme a systematic name in addition to its existing trivial name. The systematic name includes the name of the substrate or substrates in full and a word ending in "ase" indicating the nature of the reaction catalyzed. This word is either one of the six main classes of enzymes or a subdivision of one of them.

$$\underset{\text{Lactic acid}}{CH_3-\overset{H}{\underset{OH}{C}}-COOH} + NAD^+ \rightleftharpoons \underset{\text{Pyruvic acid}}{H_3C-\overset{O}{\overset{\|}{C}}-COOH} + NADH + H^+$$

Systematic name: L-lactate : NAD^+ oxidoreductase

EC Number: 1.1.1.27

Trivial name: Lactate dehydrogenase

3.5 CHEMICAL NATURE AND PROPERTIES OF ENZYMES

All the enzymes are proteins, except ribozymes, which are RNA molecules. Each enzyme has a characteristic tertiary structure and specific conformation, which is very essential for its catalytic activity. The functional unit of the enzyme is known as holoenzyme, and consists of apoenzyme, which is the protein part, and a coenzyme, which is the non-protein part, a organic molecule or a metal ion, when it is known as cofactor.

Holoenzyme (Active enzyme) ⟶ Apoenzyme (Protein part) + coenzyme (Non-protein part)

The term prosthetic group is used when the non-protein moiety is tightly (covalently) bound to the apoenzyme. The coenzyme (organic compound) can be separated from the enzyme by dialysis, while prosthetic group cannot be separated.

3.6 MONOMERIC AND OLIGOMERIC ENZYMES

The term monomeric enzyme is used if the protein contains a single polypeptide chain. For example, ribonuclease, trypsin, etc. Some of the enzymes contain more than one polypeptide chains, which may be similar or of different type, and are known as oligomeric enzymes. For example, lactate dehydrogenase contains four polypeptide chains.

There are some enzymes which contain more than one polypeptide chain and catalyze different reactions in a sequence. Only the native multienzyme complex is functionally active and not the individual polypeptide chains. For example, pyruvate dehydrogenase.

3.7 FACTORS AFFECTING ENZYME ACTIVITY

The association of enzyme and substrate is the most essential pre-requisite for the activity of an enzyme. Various factors influence the velocity of the enzyme catalyzed reactions.

3.7.1 Effect of Enzyme Concentration

As the concentration of the enzyme is increased, the velocity of the reaction also increases, provided substrate is present at very high concentration.

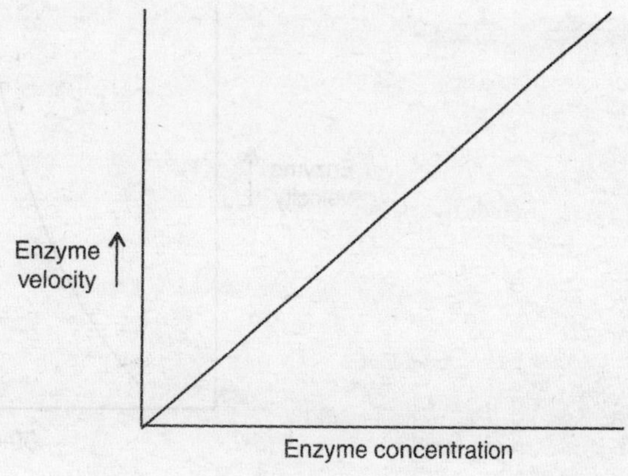

Fig. 3.2

3.7.2 Effect of Substrate Concentration

As the substrate concentration gradually increases, the velocity of enzyme catalyzed reaction also increases. However, at a very high substrate concentration, the velocity of enzyme catalyzed reaction remains constant, since all the enzyme will be saturated with the substrate, and hence any further increase in velocity cannot be achieved. A plot of velocity vs substrate concentration, gives a rectangular hyperbolic curve.

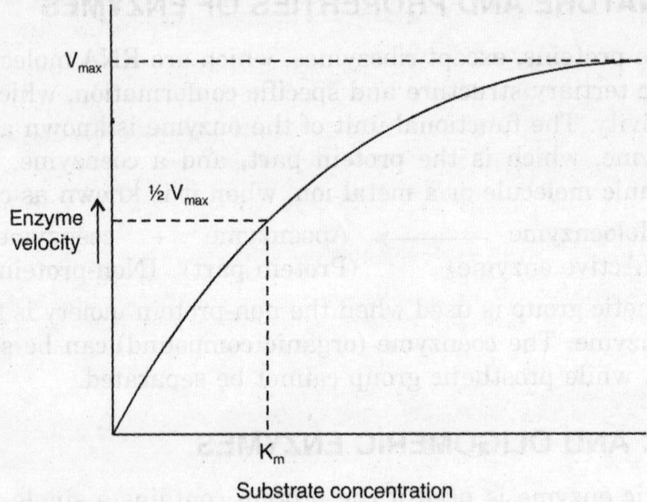

Fig. 3.3

3.7.3 Effect of Temperature

The velocity of an enzyme catalyzed reaction increases with increase in temperature up to a maximum and then declines due to the denaturation of the enzyme protein i.e., destruction of the native conformation. A bell-shaped curve is obtained when velocity is plotted against temperature.

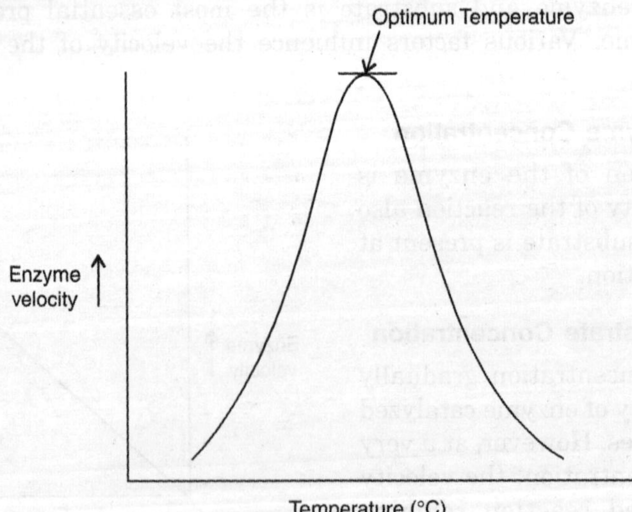

Fig. 3.4

Increase in temperature increases the activation energy and more molecular collision between enzyme and substrate occurs. This increases the rate of reaction. The optimum temperature for most enzymes lie between 35° C to 45° C. However, a few enzymes are active even at 100° C. Some plant enzymes have optimum activity around 60° C. This variable optimum temperature exhibited by different enzymes is due to the very stable structure and conformation.

The enzymes have been assigned optimal temperature based on the laboratory work. However, these temperatures have less relevance and biological significance in the living organisms.

3.7.4 Effect of pH

Hydrogen ions influence the enzyme activity by altering the ionic charges on the amino acids side chain, especially at the active site, substrate and ES complex. A gradual increase in pH (decrease in hydrogen ion concentration) increases the enzyme activity, reaches a maximum and then any further increase in pH results in decrease in enzyme activity. A plot of velocity vs pH gives a bell shaped curve.

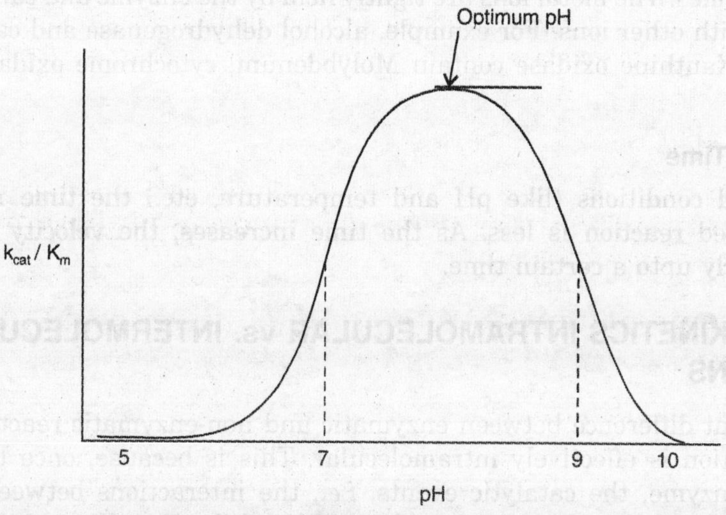

Fig. 3.5

Each enzyme has an optimum pH at which the enzyme activity is maximum. Below and above this pH, the enzyme activity is much lower and at extremes of pH, the enzyme becomes totally inactive, due to the denaturation of the enzyme.

Most of the enzymes of higher organisms show optimum activity around neutral pH of 6–8. However, some of the enzymes, such as pepsin has a optimum pH of 1–2, while acid phosphatase and alkaline phosphatase have an optimum pH of 4–5 and 10–11, respectively. Enzymes from fungi and plants are active in the pH range of 4–6.

3.7.5 Effect of Product Concentration

As enzyme catalyzed reaction proceeds, products will be formed, and it gets accumulated over a period of time. As the product concentration increases, the velocity of the reaction decreases. In the living organisms, the accumulation of product is prevented by a quick removal of products formed, since usually the product will be the substrate for the next enzyme in a metabolic pathway. However, in a metabolic pathway, the end product inhibits the activity of the first enzyme, and hence unnecessary accumulation of intermediates and products do not take place (feedback inhibition).

3.7.6 Effect of Activators

Some of the enzymes require certain inorganic metallic cations, such as Mg^{2+}, Zn^{2+}, Ca^{2+}, Co^{2+}, Cu^{2+}, Na^{+}, K^{+}, etc. for their catalytic activity. Rarely anions are required for their activity (for example, salivary amylase requires Cl^{-}). Metal ions functions as activators of enzyme activity by various mechanisms. They may combine with the substrate, form ES-metal complex, or directly participate in the reaction, or change the conformation of the enzyme.

Two categories of enzymes have been distinguished – Metal-activated enzymes and Metalloenzymes.

Metal-activated enzymes: The metal ion is loosely held by the enzyme and can be easily removed or exchanged with other ions. For example, ATPase (Mg^{2+} and Ca^{2+}).

Metalloenzymes: The metal ions are tightly held by the enzyme and cannot be removed or exchanged with other ions. For example, alcohol dehydrogenase and carboxypeptidase contain Zinc, Xanthine oxidase contain Molybdenum, cytochrome oxidase contain iron and copper.

3.7.7 Effect of Time

Under standard conditions (like pH and temperature, etc.) the time required for an enzyme catalyzed reaction is less. As the time increases, the velocity of the reaction increases linearly upto a certain time.

3.8 ENZYME KINETICS INTRAMOLECULAR vs. INTERMOLECULAR REACTIONS

A very important difference between enzymatic and non-enzymatic reactions is that the enzymatic reaction is effectively intramolecular. This is because, once the substrate is bound to the enzyme, the catalytic events, *i.e.*, the interactions between the enzyme's catalytic groups and the substrate to make and break bonds, all occur within the same molecule. This has profound effects on the entropy of the reaction, compared to the analogous reaction without enzyme catalysis. Most *intramolecular* reactions are much faster than their corresponding *intermolecular* counterparts.

$$A + B \rightarrow P \text{ and } A - B \rightarrow P$$

Kinetics are very important in the study of enzymes for many reasons: they are the reflection of the catalytic reaction, they provide important mechanistic information, they reflect the effect of pH, temperature and other environmental factors on the enzyme and catalytic reaction. They are critical in the function of the enzyme in its physiological milieu.

Many enzyme-catalyzed reactions show an initial velocity vs. [substrate] relationship of the form shown below:

$$E + S \underset{k_{-1}}{\overset{k_1}{\rightleftharpoons}} ES \xrightarrow{k_2} E + F$$

where E = enzyme, S = substrate and ES represents the enzyme-substrate complex.

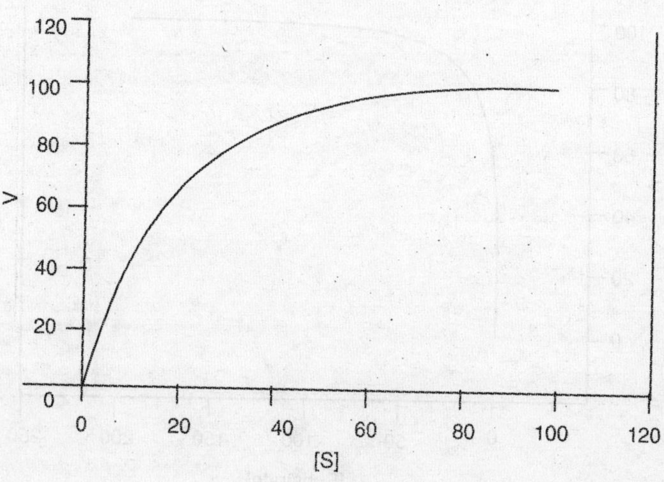

Fig. 3.6

The kinetics of the enzyme-catalyzed reaction is rather different than that of a typical chemical reaction. For example, for the reaction A → B, the uncatalyzed reaction will exhibit first-order kinetics, *i.e.*, a plot of rate (v) against [A] is linear since v = k [A],

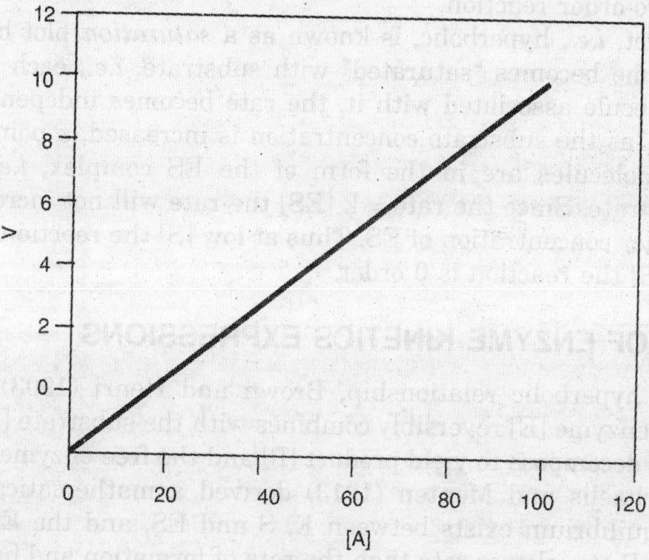

Fig. 3.7

On the other hand, enzyme-catalyzed reactions often show hyperbolic plots of v vs. [S]

$$E + S \underset{k_{-1}}{\overset{k_1}{\rightleftharpoons}} ES \xrightarrow{k_2} E + F$$

where E = enzyme, S = substrate.

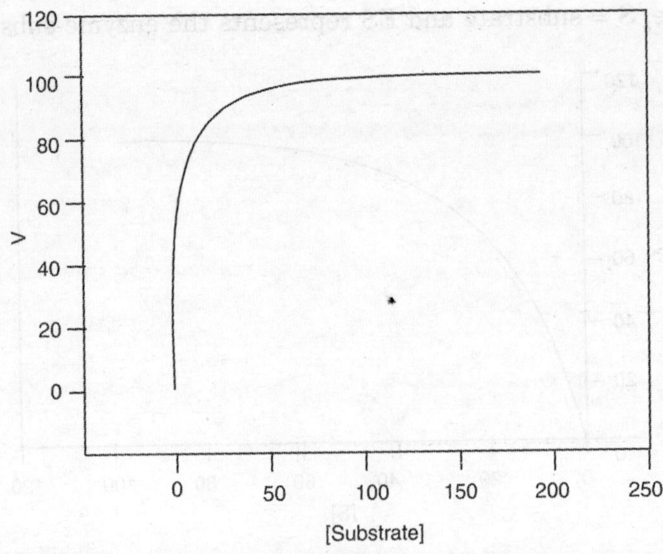

Fig. 3.8

If we start with a high substrate concentration (>mM) we observe an initial linear region which then levels off, *i.e.*, hyperbolic. Such a curve is neither 0, 1st, or 2nd order. The initial part of the curve corresponds to a first-order reaction, and the final part corresponds to a zero-order reaction.

This sort of a plot, *i.e.*, hyperbolic, is known as a *saturation* plot because it implies that when the enzyme becomes "saturated" with substrate, *i.e.*, each enzyme molecule has a substrate molecule associated with it, the rate becomes independent of substrate concentration. Thus, as the substrate concentration is increased, a point will be reached when all enzyme molecules are in the form of the ES complex, *i.e.*, the enzyme is saturated with substrate. Since the rate = k [ES] the rate will not increase further since there can be no higher concentration of ES. Thus at low [S] the reaction is first-order (v = k [S]) and at high [S] the reaction is 0 order.

3.9 DERIVATION OF ENZYME KINETICS EXPRESSIONS

To account for the hyperbolic relationship, Brown and Henri (1903) put forward the hypothesis that the enzyme [E] reversibly combines with the substrate [S] to form an [ES] complex which then decompose to yield product [P] and the free enzyme [E]. Based on the work of Henri, Michaelis and Menten (1913) derived a mathematical equation. They assumed a rapid equilibrium exists between E, S and ES, and the ES complex breaks down to form E and P at a slower rate than the rate of formation and breakdown of ES to E and S.

3.9.1 Rapid Equilibrium Assumption

This assumption is that the equilibration between E, S and ES is fast compared to the subsequent reaction of ES. *i.e.*, ES will always be in equilibrium with E and S. [ES] breaks down to form E and P at a rate slower than the rate of formation and breakdown of ES to E and S. This means that ($k_1 >> k_{-1} + k_2$ ($k_2 << k_1$ or k_{-1}).

The dissociation constant, $K_s = k_{-1}/k_1 = [E][S]/[ES]$.

$$v = k_2\,[ES] \qquad \text{...(1)}$$

The problem with this expression is that we have no way of determining the concentration of ES.

The total amount of enzyme is constant, *i.e.*, the sum of the concentration of the different enzyme-containing species is constant:

$$[E_o] = [E] + [ES]$$

$$[E] = [E_o] - [ES] \qquad \text{...(2)}$$

According to rapid equilibrium assumption, *i.e.*,

$$K_s = [E][S]/[ES] \qquad \text{...(3)}$$

Substitute [E] in the above equation

$$K_s = \{[E_o] - [ES]\}\,[S]/[ES]$$

$$K_s\,[ES] = \{[E_o] - [ES]\}\,[S]$$

$$K_s\,[ES] = ([E_o]\ [S]) - ([ES]\,[S])$$

$$K_s\,[ES] + [ES]\,[S] = [E_o]\ [S]$$

$$[ES]\,([S] + K_s) = [E_o]\,[S]$$

$$[ES] = [E_o]\,[S]/([S] + K_s)$$

[ES] governs the rate of formation of products (the overall rate of formation) according to the relationship:

$$v = k_2\,[ES] \quad \text{or} \quad [ES] = v/k_2$$

If we substitute the expression for [ES] derived above, then,

$$v = k_2\,[E_o]\,[S]/([S] + K_s) = k_{cat}\,[E_o]\,[S]/([S] + K_s)$$

where k_{cat} is the first order rate constant equal to k_2.

When the substrate concentration is very high, all the enzyme is present as the [ES] complex, and the limiting initial velocity V_{max} is reached. Under these conditions,

$$V_{max} = k_2\,[E_o]$$

Therefore, we can substitute V_{max} for $k_2\,[E_o]$ in the expression for v, and get

$$v = V_{max}\,[S]/([S] + K_s)$$

It was assumed by Michaelis and Menten that the substrate was usually present in much greater concentrations than the enzyme. Hence, the formation of the [ES] complex will result in an insignificant change in free substrate concentration. Therefore, we can substitute $[S_o]$ for [S] in the above equation.

$$v = V_{max}\,[S_o]/([S_o] + K_s)$$

This is the Michaelis – Menten equation.

3.9.2 Steady-State Approximation

The equation derived by Michaelis – Menten was modified by Briggs and Haldane and they introduced a general assumption that of steady-state. As long as $E_o << S_o$ for the majority of the reaction time, the concentration of ES will be essentially constant and the rate of change of [ES] would be negligible compared to the rate of change of [P]. Hence, once the ES complex is formed, it would be maintained at a steady-state. It would be broken down as fast as it it formed, and [ES] remain constant. The steady-state is usually established within a few milliseconds of the start of the reaction and is maintained for a few minutes until the product concentration, and hence the rate of the reverse reaction, becomes significant.

$$E + S \underset{k_{-1}}{\overset{k_1}{\rightleftharpoons}} ES \xrightarrow{k_2} E + F$$

The rate of formation of ES at any time t = k_1 [E] [S]

The rate of breakdown of ES at this time t = k_{-1} [ES] + k_2 [ES]

[ES] can breakdown to form products. According to the steady-state assumption.

$$k_1 [E] [S] = k_{-1} [ES] + K_2 [ES] = [ES] [k_{-1} + k_2]$$

Separate the constants from the variables.

$$[E] [S] / [ES] = [k_{-1} + k_2] / k_1 = K_m$$

where K_m is another constant.

The total amount of enzyme is constant, *i.e.*, the sum of the concentration of the different enzyme-containing species is constant:

$$[E_o] = [E] + [ES]$$

$$[E] = [E_o] - [ES]$$

According to steady-state assumption, *i.e.*,

$$K_m = [E][S]/[ES]$$

Substitute [E] in the above equation.

$$K_m = \{[E_o] - [ES]\} [S]/[ES]$$

$$K_m [ES] = \{[E_o] - [ES]\} [S]$$

$$K_m [ES] = ([E_o] [S]) - ([ES] [S])$$

$$K_m [ES] + [ES] [S] = [E_o] [S]$$

$$[ES] ([S] + K_m) = [E_o] [S]$$

$$[ES] = [E_o] [S]/[S] + K_m)$$

[ES] governs the rate of formation of products (the overall rate of formation) according to the relationship

$$v = k_2 [ES] \text{ or } [ES] = v/k_2$$

If we substitute the expression for [ES] derived above, then,

$$v = k_2 [E_o] [S]/([S] + K_m) = k_{cat} [E_o] [S]/([S] + K_m)$$

where k_{cat} is the first order rate constant equal to k_2.

When the substrate concentration is very high, all the enzyme is present as the [ES] complex, and the limiting initial velocity V_{max} is reached. Under these conditions,

$$V_{max} = k_2 [E_o]$$

Therefore, we can substitute V_{max} for $k_2 [E_o]$ in the expression for v, and get

$$v = V_{max} [S]/([S] + K_m)$$

It is assumed that the substrate is usually present in much greater concentrations than the enzyme. Hence, the formation of the [ES] complex will result in an insignificant change in free substrate concentration. Therefore, we can substitute $[S_o]$ for [S] in the above equation.

$$v = V_{max} [S_o]/([S_o] + K_m)$$

The above equation is same as the equation derived by Michaelis – Menten. Only the definition of the constant in the denominator has changed. Hence, this equation has retained the name Michaelis – Menten equation and K_m is called the Michaelis constant.

Let's consider what happens under certain extremes:

If S<< K_m, v becomes V_{max} $[S]/K_m$ *i.e.*, first-order, where the rate constant = $V_{ma}x/K_m$.

If S>> K_m, v becomes V_{max}, *i.e.*, zero-order (saturation).

3.10 SIGNIFICANCE OF KINETIC CONSTANTS

$\mathbf{K_{cat}}$ is a first-order rate constant corresponding to the slowest step or steps in the overall catalytic pathway. It represents the maximum number of molecules of substrate which can be converted into product per enzyme molecule per unit time (which only happens if the enzyme is "saturated" with substrate), and thus is often known as the turnover number.

$\mathbf{K_m}$ is an apparent dissociation constant, and is related to the enzyme's affinity for the substrate; it is the product of all the dissociation and equilibrium constants prior to the first irreversible step in the pathway. Often it is a close measure of the ES dissociation constant. It is defined as the substrate concentration required to produce half-maximum velocity in an enzyme catalyzed reaction. K_m value is a constant and a characteristic feature of a given enzyme. A low K_m value indicates a strong affinity between enzyme and substrate, whereas a high K_m reflects a low affinity.

$\mathbf{K_{cat}/K_m}$ is a second-order rate constant which refers to the free enzyme (not ES complex). It is also a measure of the overall efficiency of the enzyme catalysis and is known as the specificity constant.

3.11 LINEWEAVER-BURK DOUBLE RECIPROCAL PLOT

For the determination of K_m and V_{max}, the direct plot of v vs substrate concentration is not useful. Hence the michaelis – menten equation is linearized by taking the reciprocal of the MM equation. A straight line equation is obtained, which is similar to y = mx + c

$$\frac{1}{v} = \frac{K_m}{V_{max}} \times \frac{1}{S} + \frac{1}{V_{max}}$$

A plot of 1/v against 1/[S] gives a straight line. The slope of the line is K_m/V_{max} and the intercept on y-axis is $1/V_{max}$. The intercept on the x-axis gives $1/K_m$.

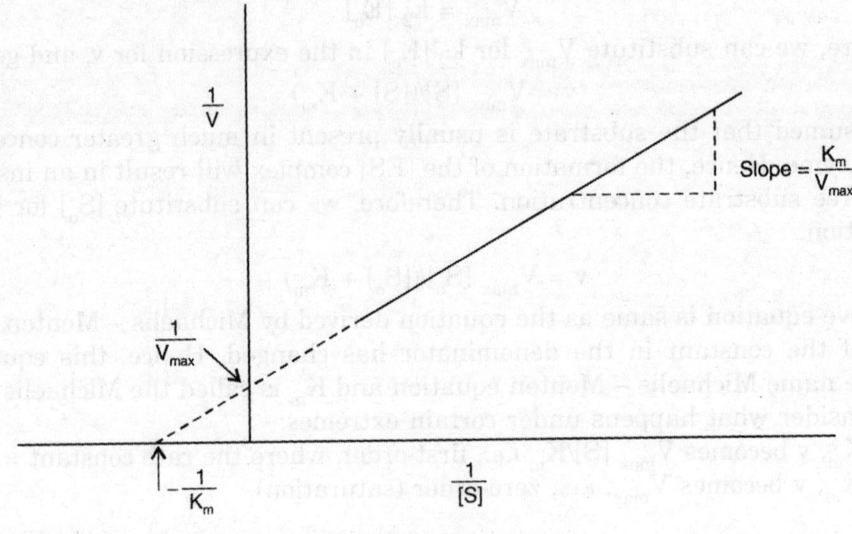

Fig. 3.9

3.12 COENZYMES AND COFACTORS

The protein part of the enzyme is not always sufficient to catalyze the reaction. Many enzymes require non-protein factors, collectively referred as cofactors for catalysis. The cofactors may be organic or inorganic in nature.

The non-protein, low molecular weight organic compound associated with the enzyme is known as coenzyme. They can be easily removed from the enzyme by dialysis. The coenzyme which is tightly bound to the enzyme and cannot be removed from the enzyme by dialysis is referred to as the prosthetic group.

The activator is referred to the inorganic cofactor (Mg^{2+}, Zn^{2+}, Ca^{2+}, Co^{2+}, Cu^{2+}, Na^{+}, K^{+}) necessary to increase the enzyme activity.

Coenzymes are regarded as the second substrate or co-substrate, since they bind to the enzyme like substrate binds to the enzyme, and undergo chemical change during the enzymatic reactions. However, they are later regenerated. In contrast, the substrate is converted to the product. Most of the coenzymes are derived from the water soluble vitamin B-complex. Coenzymes participates in various reactions involving transfer of atoms or groups like hydrogen, aldehyde, keto, amino, acyl, methyl, carbon dioxide, etc.

Coenzymes of B-complex Vitamins

Coenzyme	*Derived from Vitamin*	*Atom or group transferred*	*Enzyme*
Thiamine pyro-phosphate (TPP)	Thiamine	Aldehyde or keto	Transketolase
Flavin mono-nucleotide (FMN)	Riboflavin	Hydrogen and electron	L-amino acid oxidase

(Contd.)

Coenzyme	Derived from Vitamin	Atom or group transferred	Enzyme
Nicotinamide adenine dinucleotide (NAD^+)	Niacin	Hydrogen and electron	Lactate dehydrogenase
Lipoic acid	Lipoic acid	Hydrogen and electron	Pyruvate dehydrogenase
Pyridoxal phosphate	Pyridoxine	Amino or keto	Alanine transaminase
Coenzyme A (CoA)	Pantothenic acid	Acyl	Thiokinase
Tetrahydrofolate (THF)	Folic acid	One carbon	Formyl transferase
Biotin	Biotin	CO_2	Pyruvate carboxylase

Coenzymes not Related to B-complex Vitamins

Not all coenzymes are vitamin derivatives. There are some other organic molecules, which have no relation to vitamins, but function as coenzymes. They are non-vitamin coenzymes. For example, ATP, CDP, UDP, S – adenosylmethionine (SAM), etc.

Coenzyme	Biochemical function
ATP	Donates phosphate, adenosine monophosphate moiety (AMP)
CDP	Required in phospholipids synthesis as carrier of choline and ethanolamine
UDP	Required for glycogen synthesis as carrier of glucose
SAM	Donates methyl group in biosynthetic reactions

Nucleotide Coenzymes

Some of the coenzymes possess nitrogenous base, sugar, and phosphate. Hence, such coenzymes are regarded as nucleotide coenzymes. For example, NAD^+, $NADP^+$, FMN, FAD, UDPG, Coenzyme A, etc.

4 Carbohydrates

4.1 INTRODUCTION

Carbohydrates are the most abundant renewable source of energy. They are synthesised in plants by the process of photosynthesis, and are utilised by man as source of energy. Carbohydrates also play an important role in food by providing desirable texture.

Carbohydrates, as the name indicates, is made of both carbon and hydrogen. Generally, carbohydrates are defined as polyhydroxy aldehydes or ketones and their derivatives or compounds which yield these derivatives upon hydrolysis. Many of them have the empirical formula $(CH_2O)_n$, indicating that they are hydrates of carbon.

4.2 FUNCTION

1. Carbohydrates have both structural and functional role to play in every living cell. For example, polymeric carbohydrates of cell walls of plant and bacteria – cellulose and peptidoglycan.
2. Simpler carbohydrates are utilised for energy production. For example, glucose.
3. Carbohydrates are used for the synthesis of non-carbohydrate constituents.
4. Some carbohydrates have highly specific function. For example, ribose and deoxyribose in nucleic acids, galactose in certain lipids, and mannose in glycoproteins.
5. Carbohydrates confer biological specificity on the surface of animal cells.

4.3 CLASSIFICATION

Carbohydrates are classified as monosaccharides, oligosaccharides and polysaccharides.

4.3.1 Monosaccharides

Monosaccharides are the simple sugars that cannot be hydrolysed to simpler forms. About 200 different types of monosaccharides are found in nature. Glucose (dextrose or blood sugar) is the most prevalent monosaccharide while fructose (obtained from fruits and honey) is the sweetest of all monosaccharides, which is slowly absorbed by the gut and converted to glucose in the liver.

Depending on the number of carbon atoms they contain they are sub-divided into trioses, tetroses, pentoses, hexoses, heptuloses, etc., containing 3, 4, 5, 6, 7 etc. carbon atoms, respectively. Further, depending on whether they contain aldehyde or ketone group, they are further classified as aldotriose, ketotriose, aldotetrose, ketotetrose, aldopentose, ketopentose, aldohexose, ketohexose, etc.

Aldotriose (D-glyceraldehyde)	Aldotetrose (D-Erythrose)	Aldopentose (D-Ribose)	Ketotriose (Dihydroxyacetone)	Ketotetrose (D-Erythrulose)	Ketopentose (D-Ribulose)
CHO	CHO	CHO	CH_2OH	CH_2OH	CH_2OH
HCOH	HCOH	HCOH	C=O	C=O	C=O
CH_2OH	HCOH	HCOH	CH_2OH	HCOH	HCOH
	CH_2OH	HCOH		CH_2OH	HCOH
		CH_2OH			CH_2OH

Glucose is the most abundant monosaccharide used as major fuel for deriving energy in most organisms. It is also the basic building block of the most abundant polysaccharide, such as starch, glycogen and cellulose.

4.3.2 Oligosaccharides

Oligosaccharides (oligo means few) contain 2 – 10 monosaccharide units joined by glycosidic linkage. The major type of oligosaccharides are the disaccharides, such as sucrose (glucose + fructose), lactose (glucose + galactose), and maltose (glucose + glucose). Sucrose is the most common dietary disaccharide occurring in – beet root and cane sugar, brown sugar, sorghum, maple syrup and honey. It contributes up to 25% of the total quantity of ingested cals in U.S. Lactose is only found in milk and is often called "milk sugar". Maltose occurs in malt products and in germinating cereal (is negligible in an average person's diet).

4.3.3 Polysaccharides

Most of the polysaccharides found in nature occur as polysaccharides of high molecular weight, formed by the condensation of individual monosaccharide units or amino sugars or uronic acid derived from them, in a long linear or branched chains. Depending on the type of monosaccharides present, they are classified as homopolysaccharides (starch, glycogen, cellulose), which contain only one type of recurring monosaccharide units, and heteropolysaccharides (galactomannans, galactoglucomannans), which contain two or more types of recurring monosaccharide units.

Homopolysaccharides containing only one type of monosaccharides such as xylose, arabinose, glucose, mannose, galactose, are known as xylans, arabinans, glucans, mannans, galactans, respectively. Homopolysaccharides are further classified as pentosans and hexosans, depending on the type of hydrolytic products formed.

4.4 ROLE OF CARBOHYDRATE IN THE BODY

1. Energy Source
2. Protein Sparing

3. Metabolic Primer
4. Fuel for the Central Nervous System

4.4.1 Energy Source

The main function of carbohydrate is to serve as an energy source for the body. Energy liberated from the catabolism of glucose and glycogen is used to power muscle contraction and other biological processes. Once the capacity of the cell for glycogen storage is reached, the excess sugars are converted and stored as fat, even if the diet is low in fat.

4.4.2 Protein Sparing

When carbohydrate reserves are reduced, metabolic pathways exist for the synthesis of glucose from protein and the glycerol portion of a fat molecule. However, through this mechanism, the kidneys can be damaged as they must handle an increased load of excretion of nitrogen-containing byproducts of protein breakdown. Therefore, glycogen depletion which occurs in athletes, quite often (especially aerobic athletes) can result in a loss of muscle mass and strength.

4.4.3 Metabolic Primer

Carbohydrates act as a primer for fat metabolism. If insufficient carbohydrate metabolism exists, either through limitation in glucose transport into the cell (diabetes), or depletion of glycogen through improper diet or prolonged exercise, the body begins to mobilise fat to a greater extent than it can use. This results in incomplete fat metabolism and the accumulation of acids by-products called ketone bodies. This may result in harmful increase in the acidity of body fluids, a condition called acidosis or more specifically with regard to fat breakdown Ketosis.

4.4.4 Fuel for the Central Nervous System

Under normal conditions and in short-term starvation, the brain uses blood glucose almost exclusively as a fuel and essentially has no stored supply of this nutrient. The symptoms of a modest reduction in blood glucose (hypoglycemia) include feelings of weakness, hunger, and dizziness. This condition impairs exercise performance and may partially explain the fatigue associated with prolonged exercise. Sustained and profound hypoglycemia can result in irreversible brain damage.

4.5 CARBOHYDRATES AS A SOURCE OF ENERGY

Starch is the main source of carbohydrates in the diet for most of the humans. Digestion begins in the mouth, where salivary α-amylase hydrolyses the glycosidic linkages of starch and produces short polysaccharide fragments or oligosaccharides. In the stomach, salivary α-amylase is inactivated by the low pH. However, a second form of α-amylase secreted by pancreas into the small intestine continues the degradation process. Pancreatic amylase produces mainly maltose and oligosaccharides called dextrins, which are fragments of the amylopectin containing α-1, 6 branch points. Maltose and dextrins are hydrolysed to monosaccharides before entering the cells. They are degraded by enzymes attached to outer surface of the intestinal epithelial cells.

$$\text{Dextrin} + nH_2O \xrightarrow[\text{Dextrinase}]{} n\,\text{D-glucose}$$

$$\text{Maltose} + H_2O \xrightarrow[\text{Maltase}]{} 2\,\text{D-glucose}$$

The monosaccharides so formed are transported into the epithelial cells, which then pass into the blood to be carried to various tissues, where they are phosphorylated and funneled into the glycolytic pathway.

Dietary glycogen has essentially the same structure as starch, and its digestion proceeds by the same pathway.

4.5.1 Other Monosaccharides enter the Glycolytic Pathway at Several Points

In most organisms, hexoses (such as mannose, fructose, galactose) other than glucose can undergo glycolysis after conversion to a phosphorylated derivative.

$$\text{Lactose} + H_2O \xrightarrow[\text{Lactase}]{} \text{D-glucose} + \text{D-galactose}$$

$$\text{Sucrose} + H_2O \xrightarrow[\text{Sucrase}]{} \text{D-glucose} + \text{D-fructose}$$

$$\text{Trehalose} + H_2O \xrightarrow[\text{Trehalase}]{} \text{2D-glucose}$$

4.5.1.1 Mannose

D-mannose produced during the digestion of various polysaccharides and glycoproteins of foods, can be phosphorylated at C-6 position by hexokinases to form mannose-6-phosphate, which is isomerised by phosphomannose isomerase to yield fructose-6-phosphate, an intermediate of glycolysis.

$$\text{Mannose} + \text{ATP} \xrightarrow[\text{Hexokinase}]{} \text{Mannose-6-phosphate} + \text{ADP}$$

$$\text{Mannose-6-phosphate} \xrightarrow[\text{Phosphomannose isomerase}]{} \text{Fructose-6-phosphate}$$

4.5.1.2 Fructose

D-fructose present in the free form in many fruits and also produced by hydrolysis of sucrose in the small intestine of vertebrates, is phosphorylated by hexokinase to form fructose-6-phosphate. This is the major pathway for fructose to enter glycolysis in the muscles and kidney.

$$\text{Fructose} + \text{ATP} \xrightarrow[\text{Hexokinase}]{} \text{Fructose-6-phosphate} + \text{ADP}$$

However, in the liver, fructose enters into the glycolytic pathway by a different route. The enzyme fructokinase in the liver catalyzes the phosphorylation of fructose at C-1 rather than C-6 position.

$$\text{Fructose} + \text{ATP} \xrightarrow[\text{Fructokinase}]{} \text{Fructose-1-phosphate} + \text{ADP}$$

Fructose-1-phosphate is then cleaved by the enzyme fructose-1-phosphate aldolase to form glyceraldehyde and dihydroxyacetone phosphate.

$$\text{Fructose-1-phosphate} \xrightarrow[\text{Aldolase}]{} \text{Glyceraldehyde} + \text{Dihydroxyacetone phosphate}$$

Dihydroxyacetone phosphate is isomerised to glyceraldehyde-3-phosphate by the glycolytic enzyme triose phosphate isomerase.

$$\text{Dihydroxyacetone phosphate} \xrightleftharpoons[\text{Isomerase}]{} \text{Glyceraldehyde-3-phosphate}$$

Glyceraldehyde is phosphorylated by the enzyme triose kinase to form Glyceraldehyde-3-phosphate.

$$\text{Glyceraldehyde} + \text{ATP} \xrightarrow[\text{Triose kinase}]{} \text{Glyceraldehyde-3-phosphate} + \text{ADP}$$

Thus, both the products of fructose enter the glycolytic pathway as glyceraldehydes-3-phosphate.

4.5.1.3 Galactose

D-galactose is the hydrolytic product of lactose. Galactose is first phosphorylated by the enzyme galactokinase at C-1 by ATP to form galactose-1-phosphate.

$$\text{Galactose} + \text{ATP} \xrightarrow[\text{Galactokinase}]{} \text{Galactose-1-phosphate} + \text{ADP}$$

Galactose-1-phosphate is then converted to its epimer at C-4, by a set of reactions to form glucose-1-phosphate.

$$\text{Galactose-1-phosphate} + \text{UDP-glu cos e} \rightarrow \text{Glucose-1-phosphate} + \text{UDP-galactose}$$

$$\text{UDP-galactose} \rightleftharpoons \text{UDP-glucose}$$

UDP-glucose is recycled through another round of the same reaction. The net effect of this cycle is the conversion of galactose-1-phosphate to glucose-1-phosphate. There is no net production or conversion of UDP-galactose or UDP-glucose.

4.6 STEREOCHEMISTRY OF MONOSACCHARIDES

All carbohydrates, except dihydroxyacetone, contain one or more asymmetric carbon atoms and thus they are called chiral molecules and are optically active. They are capable of existing in two forms that are non-superimposable mirror images of each other. They can exist in right-handed and left-handed forms. They are called chiral compounds. The phenomenon is known as chirality. The two chiral forms are known as enantiomers. Enantiomers have the same chemical properties, but certain of their physical properties and essentially all of their physiological properties are different. They rotate the plane polarised light to equal extent, but in the opposite direction.

The presence of asymmetric carbon atom confers optical activity to the compound. When a plane polarised light is passed through a solution of an optical isomer, it might rotate the light to the right (+) or to the left (–) and are called as dextrorotatory or levorotatory, respectively.

The monosaccharides belong to either D-series or L-series, since their configurations are related to three carbon sugar, D-glyceraldehyde. The orientation of the -OH and -H groups around the asymmetric carbon atom farthest removed from the carbonyl carbon atom determines whether the sugar belongs to D-series or L-series.

CHO

H ◁ C ◁ OH

CH$_2$OH

D-glyceraldehyde

CHO

HO ◁ C ◁ H

CH$_2$OH

L-glyceraldehyde

Non-superimposable mirror images

The absolute configuration of an isomer is designated by the symbols D- and L-. The structure with –OH group on the right when –CHO group is at the top is designated as D-glyceraldehyde, and the structure with –OH group on the left when –CHO group is at the top is designated L-glyceraldehyde. A compound may be either D (+), D (–), L (+), or L (–).

Racemic Mixture: When equal amounts of D- and L- isomers are present, then it results in a racemic mixture (DL–mixture), and do not exhibit optical activity, since the dextro- and levorotatory activities cancel each other.

All sugars, whose configuration around the penultimate carbon is similar to D-glyceraldehyde are designated as D-compounds and their mirror images as L-compounds.

If a molecule has more than one asymmetric carbon atom, it can exist in 2^n forms, where n = number of asymmetric carbon atom. Thus, aldotriose (such as glyceraldehydes) having only one asymmetric carbon atom exist in $2^1 = 2$ different stereoisomeric forms. Similarly, aldohexose having four asymmetric carbon atoms exist in $2^4 = 16$ different sterioisomeric forms, while ketohexose having three asymmetric carbon atoms exist in $2^3 = 8$ different stereoisomeric forms.

The nomenclature for molecules with two asymmetric carbon atoms (four carbon sugars) is erythrose and threose. If two similar groups are on the same side, they are called as erythro-form, while if they are on the opposite side, they are called as threo-form.

Epimers: Two sugars, which differ only in the configuration around one specific carbon atom are called epimers of each other. Thus, D-glucose and D-mannose, are epimers with respect to carbon atom 2, and D-glucose and D-galactose, are epimers with respect to carbon atom 4.

D-Galactose	D-Glucose	D-Mannose
O=C–H	O=C–H	O=C–H
H—C—OH	H—C—OH	HO—C—H
HO—C—H	HO—C—H	HO—C—H
HO—C—H	H—C—OH	H—C—OH
H—C—OH	H—C—OH	H—C—OH
CH$_2$OH	CH$_2$OH	CH$_2$OH

4.7 FISCHER PROJECTION FORMULA

The structural and steriochemical relationship between sugars was established by Emil Fischer, and he was honoured by the award of Nobel Prize in 1901. To represent the 3D-structure of the molecules in 2D–, Emil Fischer's projection formula is used. The molecule is placed with the asymmetric carbon atom in the plane of the projection. The groups on top and bottom of the asymmetric carbon atoms and those in a plane behind it are also written along the straight line. The groups on either side of the asymmetric carbon atom and in a plane in front of it are written on opposite sides of the straight line.

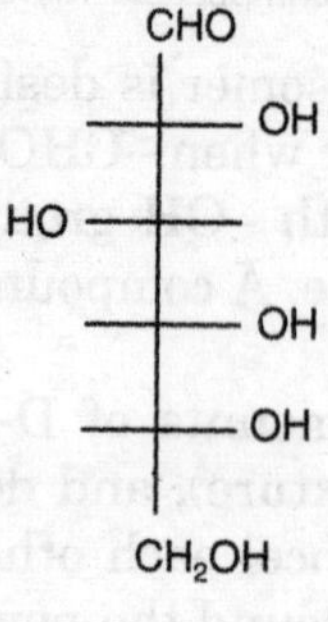

CHO
— OH
HO —
— OH
— OH
CH_2OH

D-Glucose

4.8 HAWORTH RING STRUCTURE

The Fischer projection formula for the sugars suggests that the aldehyde group and the alcohol group of C5 can readily approach each other. It is also possible for the molecule to twist in such a way that the alcohol group of C4 is brought close to the aldehyde. Similarly, the keto group of a keto sugar can approach the alcohol group of either C6 or C5. The close proximity of an aldehyde or keto group and an alcohol group in the sugars facilitate their reaction to give cyclic hemiacetal or cyclic hemiketal.

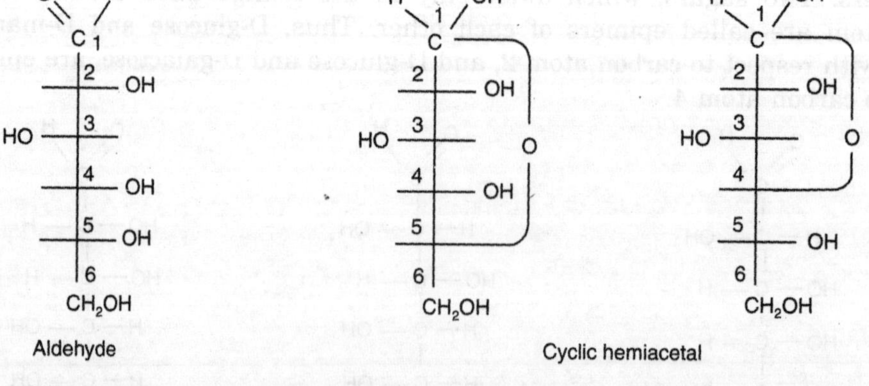

O H / C 1 / 2 — OH / HO — 3 / 4 — OH / 5 — OH / 6 / CH_2OH

Aldehyde

H OH / C 1 / 2 — OH / HO — 3 / O / 4 — OH / 5 / 6 / CH_2OH

H OH / C 1 / 2 — OH / HO — 3 / O / 4 / 5 — OH / 6 / CH_2OH

Cyclic hemiacetal

This reaction does not involve the loss or gain of any atoms by the sugar molecule. Hence, the aldehyde and cyclic forms are isomers. The sugars exist in either the aldehyde or hemiacetal form. The cyclic forms are the most important forms of the sugar molecules.

Glucose exists in the crystalline state or in water solution, almost entirely as one or another hemiacetal but not as free aldehyde. The sugar displays many reactions typical of aldehyde because hemiacetal formation is an equilibrium process, so that small amounts of the free aldehyde are always present in the solution.

The formation of the cyclic forms introduces two new structural problems into the chemistry of the sugars.

The first is the size of the ring. A cyclic hemiacetal can involve either C4 or C5. In the first case, the ring contains 4 carbon and 1 oxygen atom, and is called as furanose ring. In the second case, the ring contains 5 carbon and 1 oxygen atom, and is called as pyranose ring.

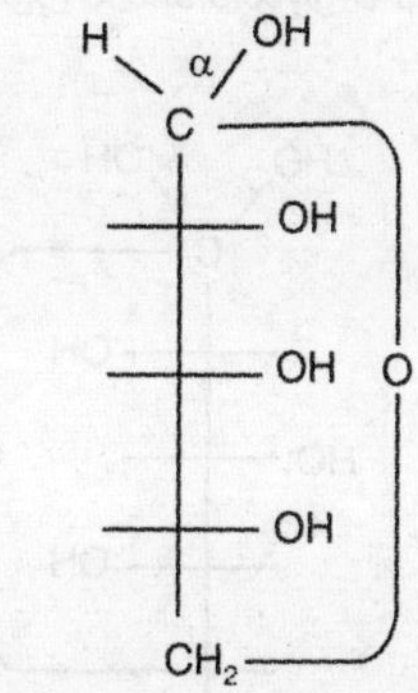

α-D-Ribopyranose

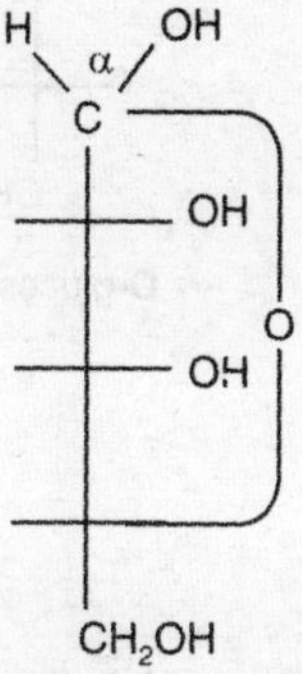

α-D-Ribofuranose

Furan

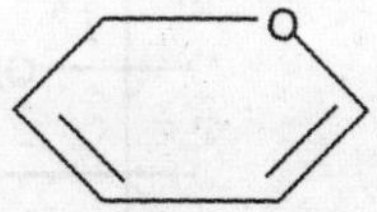

Pyran

The second variation is in the stereochemistry of hemiacetal or hemiketal — a new asymmetric carbon atom is formed. This carbon atom is called as anomeric carbon atom. If the –OH group of the hemiacetal or hemiketal is on the same side of the ring, then it is called α-, and if it is on the opposite side of the ring it is called as β-. Due to the reversible nature of hemiacetal or hemiketal formation, and also because either furanose, pyranose, α-, or β- isomers can be formed, a solution of a sugar actually contains all the possible mixtures or forms in a freely reversible equilibrium. Thus, if any one of the isomer or anomer is dissolved in water, an equilibrium mixture of all the forms will be formed. α-D-glucose in an aqueous solution exists as an equilibrium mixture of five compounds (α-D-glucopyranose, β-D-glucopyranose, α-D-glucofuranose, β-D-glucofuranose, and free aldehyde).

The interconversion between the two anomers is due to mutarotation. As a result of mutarotation, an aqueous solution of either pure α-D-glucose or β-D-glucose, will approach an equilibrium mixture containing the five compounds.

Mutarotation: The α- and β- anomers of glucose have different optical rotations. The specific optical rotation of a freshly prepared α-D-glucose in water is +112.2° which gradually changes and attains an equilibrium with a constant value of +52.7°. In the presence of alkali, the decrease in optical rotation is rapid. The optical rotation of β-D-glucose is +18.7°. Mutarotation is defined as the change in the specific optical rotation representing the interconversion of α- and β-forms of D-glucose to an equilibrium mixture. The equilibrium mixture contains 62% β-anomer and 37% of α-anomer, with a total 1% of furanose and open chain forms.

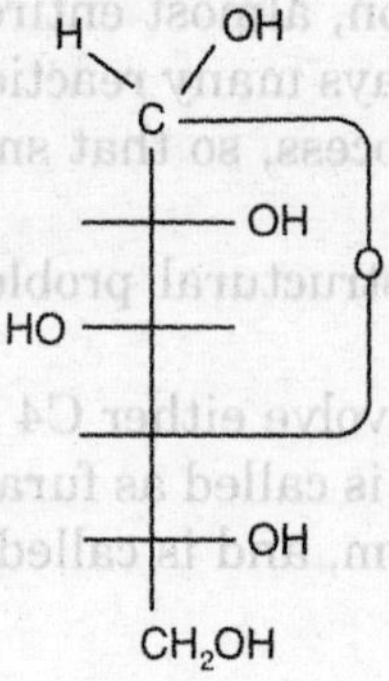

α-D-glucofuranose (0.5%)

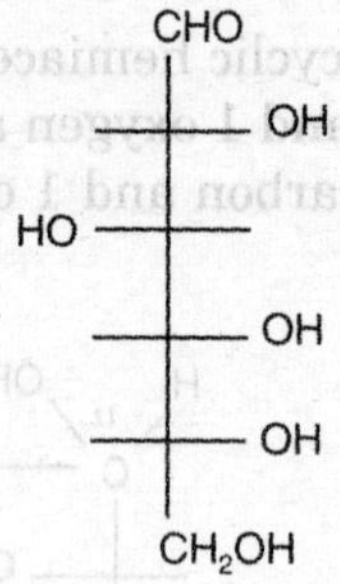

D-glucose (0.003%)

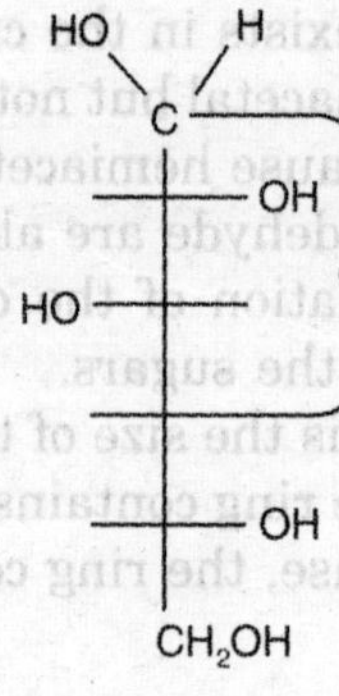

β-D-glycofuranose (0.5%)

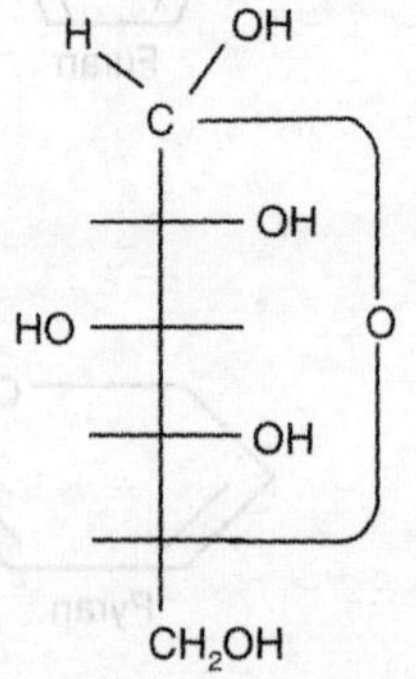

α-D-glucopyranose (37%)

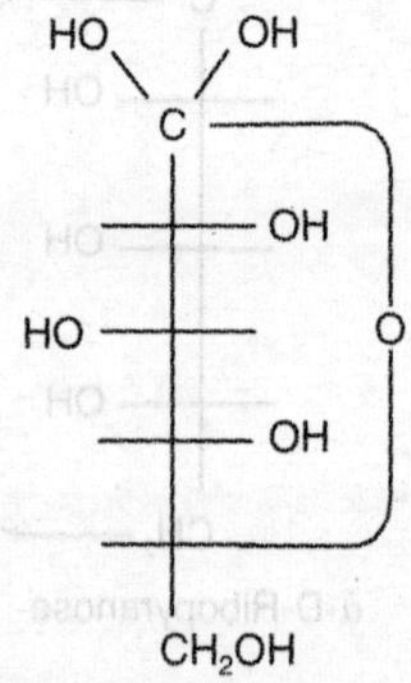

β-D-glucopyranose (62%)

4.9 CONVERSION OF FISCHER PROJECTION TO HAWORTH FORM

When the rings are viewed, the attached groups will be either above or below the plane of the ring.

1. Any group to the right of the carbon chain is written down. Those to the left are written up (*i.e.*, above the plane of the ring).
2. – OH group on the asymmetric carbon atom in D-series is down, if α-or up if it is β –

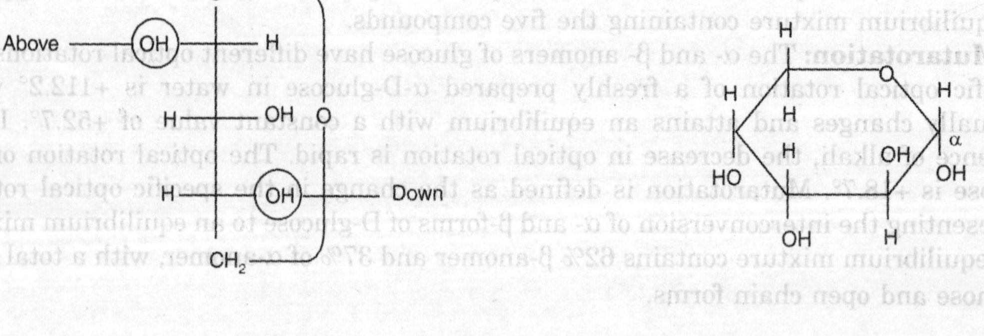

3. **When there are more carbon atoms in the sugar than are involved in ring formation, the rule is that if the ring is to right, the extra carbon (-CH_2OH) will be above the plane *i.e.*, up, and if the ring is to left, the extra carbon (-CH_2OH) will be below the plane *i.e.*, down.**

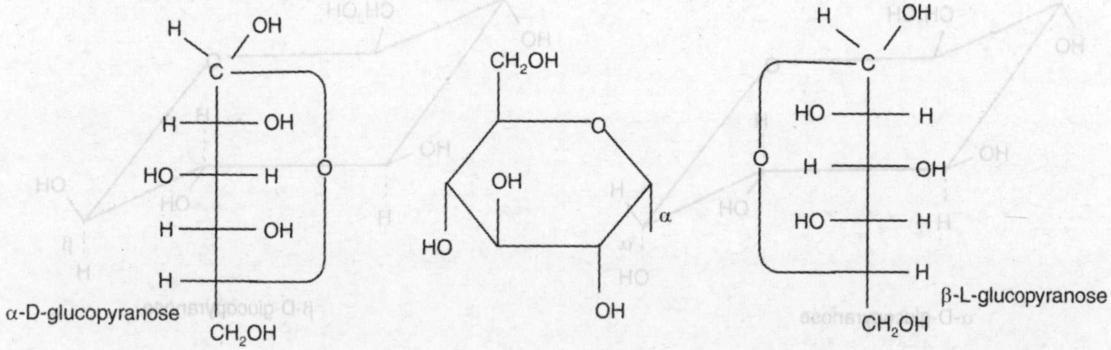

α-D-glucopyranose

β-L-glucopyranose

Ring to the right → CH_2OH up

Ring to the left → CH_2OH below

4. **In the case of a ketohexose, in the furanose form, the position of the sixth carbon atom will be determined as above, while the –OH group on first carbon atom will be down or up, depending upon whether the sugar is α- or β-form, respectively.**

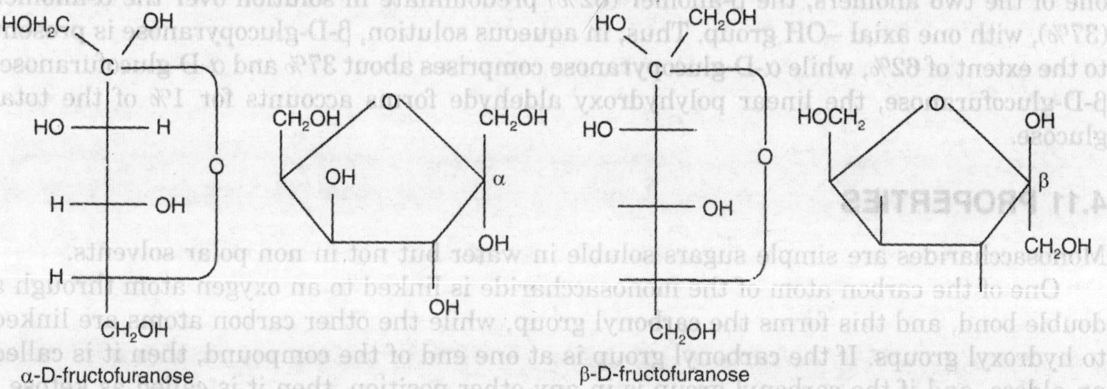

α-D-fructofuranose

β-D-fructofuranose

4.10 CONFORMATION

Although, the Howarth formulae indicate the true structure of sugars, they do not take into account the valence angles and hence do not represent the actual conformations. Six membered rings can exist in either the boat or chair form. The C-O-C bond angle of hemiacetal ring (111°) in glucose is similar to that of the C-C-C bond angle in cyclohexane (109°). The pyranose ring of glucose rather than forming a true planar structure, is puckered or twisted in much the same way as cyclohexane. Cyclohexane can occur in two conformations – the chair conformation which is more stable than the boat conformation.

The chair conformation of glucose minimises the torsional strain and further, the conformational structure in which a maximum number of bulky substituents or groups (such as –OH and $-CH_2OH$) are equatorial rather than axial to an axis passing through the ring is preferred.

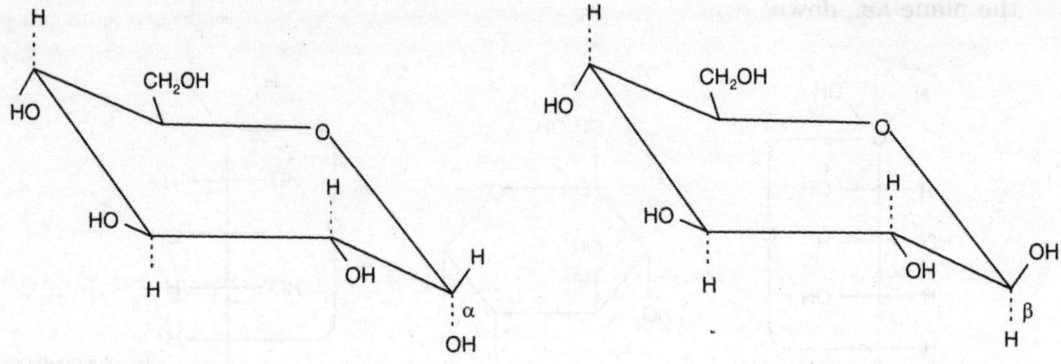

α-D-glucopyranose

β-D-glucopyranose

β-D-glucopyranose is present in a conformation in which all bulky substituents are equatorial (perpendicular to an axis passing through the plane of the ring). This conformation is thermodynamically more stable than that in which the –OH group and $-CH_2OH$ groups are axial (parallel to the axis). α-D-glucopyranose can have a conformation in which all the bulky groups except the anomeric –OH group are equatorial. Therefore, one of the two anomers, the β-anomer (62%) predominate in solution over the α-anomer (37%), with one axial –OH group. Thus, in aqueous solution, β-D-glucopyranose is present to the extent of 62%, while α-D-glucopyranose comprises about 37% and α-D-glucofuranose, β-D-glucofuranose, the linear polyhydroxy aldehyde forms accounts for 1% of the total glucose.

4.11 PROPERTIES

Monosaccharides are simple sugars soluble in water but not in non polar solvents.

One of the carbon atom of the monosaccharide is linked to an oxygen atom through a double bond, and this forms the carbonyl group, while the other carbon atoms are linked to hydroxyl groups. If the carbonyl group is at one end of the compound, then it is called an aldose, and if the carbonyl group is in any other position, then it is called as ketose.

Aldose

Ketose

A hemiacetal is formed by the action of an aldehyde with an alcohol in the presence of an acid catalyst.

$$R-\overset{O}{\overset{\|}{C}}-H + R'-OH \underset{}{\overset{H^+}{\rightleftharpoons}} R-\underset{OR'}{\overset{OH}{C}}-H \overset{H^+}{\rightleftharpoons} R-\underset{OR'}{\overset{OR'}{C}}-H$$

Hemiacetal — Acetal

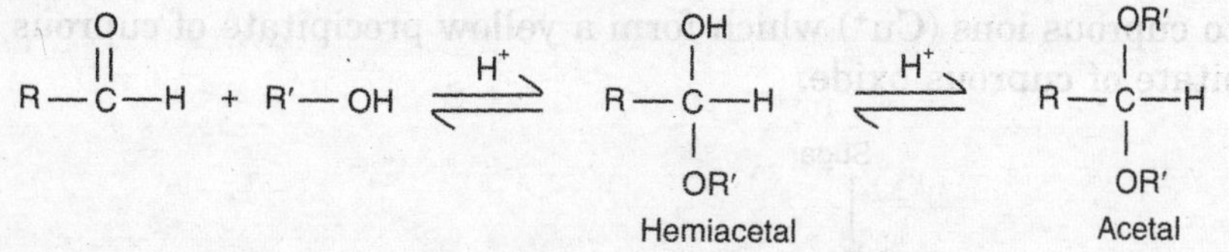

Similarly, a hemiketal is formed by the action of a ketone with alcohol in presence of an acid catalyst.

$$R-\overset{O}{\overset{\|}{C}}-R' + R''-OH \overset{H^+}{\rightleftharpoons} R-\underset{OR''}{\overset{OH}{C}}-R' \overset{H^+}{\rightleftharpoons} R-\underset{OR''}{\overset{OR''}{C}}-R'$$

Hemiketal — Ketal (Acetal)

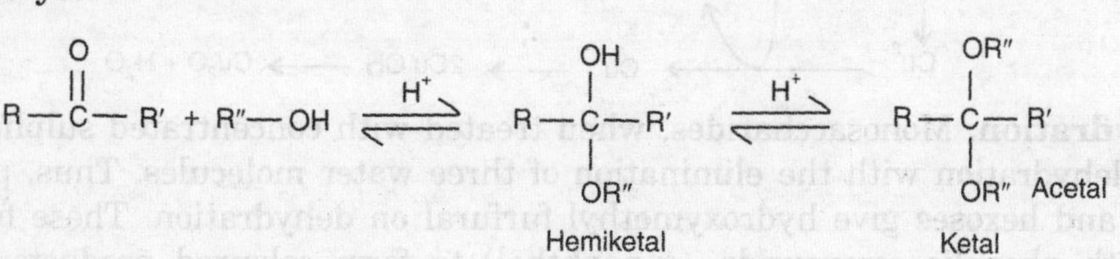

Carbohydrates in most part exist only in the form of internal hemiacetals or hemiketals. For example, if the crystalline form of D-ribose is dissolved in a suitable solvent, a very small proportion of the aldehyde form exists, which by intramolecular hemiacetal formation gives rise to four other structures.

Tautomerisation or Enolisation: Sugars containing anomeric carbon atom undergoes tautomerisation in alkaline solutions. The process involving the shifting of a hydrogen atom from one carbon atom to another to produce enediols is known as tautomerisation. When glucose is present in alkaline solution for several hours, it undergoes isomerisation to form D-fructose and D-mannose, via the formation of a common intermediate enediol. The enediols are highly reactive, hence sugars in alkaline solution act as powerful reducing agents.

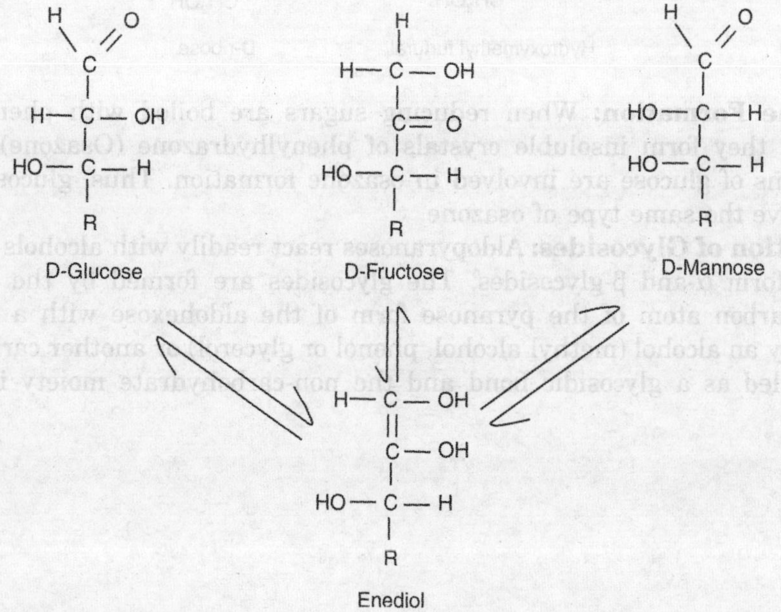

D-Glucose: H–C=O; H–C–OH; HO–C–H; R

D-Fructose: H–C(H)–OH; C=O; HO–C–H; R

D-Mannose: H–C=O; HO–C–H; HO–C–H; R

Enediol: H–C–OH; ‖ C–OH; HO–C–H; R

Reducing Properties: The sugars are classified as reducing or non-reducing sugars. This property is attributed to the free aldehyde or keto group of anomeric carbon atom. The enediol forms produced in alkaline solutions reduce cupric ions (Cu^{2+}) of copper sulphate to cuprous ions (Cu^{+}) which form a yellow precipitate of cuprous hydroxide or a red precipitate of cuprous oxide.

Sugar → Enediol → Sugar acid

Cu SO_4 → Cu^{2+} → Cu^{+} → 2Cu OH → $Cu_2O + H_2O$

Dehydration: Monosaccharides, when treated with concentrated sulphuric acid undergoes dehydration with the elimination of three water molecules. Thus, pentoses give furfural, and hexoses give hydroxymethyl furfural on dehydration. These furfurals condense with phenolic compounds, (α-naphthol) to form coloured products. This is the chemical basis of the Molisch test.

D-glucose —(conc. H_2SO_4, $3H_2O$)→ Hydroxymethyl furfural

D-ribose —(conc. H_2SO_4, $3H_2O$)→ Furfural

Osazone Formation: When reducing sugars are boiled with phenylhydrazine in acetic acid, they form insoluble crystals of phenylhydrazone (Osazone). The first two carbon atoms of glucose are involved in osazone formation. Thus, glucose, fructose and mannose give the same type of osazone.

Formation of Glycosides: Aldopyranoses react readily with alcohols in the presence of acid to form α-and β-glycosides. The glycosides are formed by the reaction of the anomeric carbon atom of the pyranose form of the aldohexose with a hydroxyl group furnished by an alcohol (methyl alcohol, phenol or glycerol) or another carbohydrate. This bond is called as a glycosidic bond and the non-carbohydrate moiety is called as the aglycone.

H—C(=O) | H—C—OH | R (Glucose) + $H_2NNH-C_6H_5$ (Phenylhydrazine) → Glucohydrazone: $H-C=N-NH-C_6H_5$ | H—C—OH | RH; + $H_2N-NH-C_6H_5$ → Glucosazone: $H-C=N-NH-C_6H_5$ | $H-C=N-NH-C_6H_5$ | R

D-glucose reacts with methyl alcohol in the presence of acid to yield methyl α-D-glycopyranoside and methyl β-D-glucopyranoside with a specific optical rotation of +158.9° and –34.2°, respectively.

CH_2OH, O, OCH_3 — Methyl-αD-glucopyranose

CH_2OH, O, OCH_3 — Methyl-βD-glucopyranose

Glycosidic linkage can also be formed by the reaction of the anomeric carbon of a monosaccharide with a hydroxyl group of another monosaccharide to yield a disaccharide. Oligosaccharides and polysaccharides are chain of monosaccharides joined by glycosidic linkages. The glycosidic linkage is stable to bases, but is hydrolysed by boiling with acid to yield the free monosaccharide and free alcohol.

CH_2OH, O, α, O, CH_2OH, O — Maltose

CH_2OH, O, β, O, CH_2OH, O — Lactose

The nomenclature of glycosidic bonds is based on the linkages between the carbon atoms and the nature of the anomeric carbon (α, or β-). For example, maltose, formed by a glycosidic bond between C1 of α-glucose of one sugar and C4 of another glucose, is named as α (1 → 4) glycosidic bond. Similarly, lactose, formed by a glycosidic bond between C1 of β – galactose and C4 of glucose, is named as β (1 → 4) glycosidic bond.

Aldoses and ketoses react with amines to form N-glycosidic bond. In nucleotides and nucleic acids, ring nitrogen atom of purine and pyrimidine bases form N-glycosidic linkages with C1 of β-D ribose or β-D-2-deoxyribose.

NH_2 4 5 3 N 2 6 O 1 N $HOCH_2$ O

N_1-(β-D-ribofuranosyl) Cytosine

4.12 DERIVATIVES OF MONOSACCHARIDES

There are several derivatives of monosaccharides, such as sugar alcohols, sugar acids, amino sugars and deoxy sugars.

Sugar Alcohols (Polyols): The carbonyl group of monosaccharides on reduction yields the corresponding sugar alcohols. Monosaccharides, when treated with reducing agents such as sodium amalgam, the aldehyde or keto group is reduced to the corresponding alcohol. Thus, D-glucose, D-galactose, D-mannose, and D-ribose are reduced to the corresponding alcohol D-sorbitol, D-dulcitol, (D-galactitol), D-mannitol, D-ribitol, respectively. Two other important sugar alcohols which occur in nature are glycerol and inositol.

Ribitol is a constituent of flavin coenzyme, glycerol and myo-inositol are components of lipids. Xylitol, is a sweetener used in sugarless gums and candies.

Sugar Acids: There are three important types of sugar acids — aldonic, uronic and aldaric acids. Depending on the oxidising agent used, the terminal aldehyde (or keto) or the terminal alcohol or both the groups may be oxidised to form sugar acids.

1. Oxidation of aldoses at the aldehyde carbon (–CHO) atom by weak oxidising agents like sodium hypoiodite results in the formation of the corresponding carboxylic acids, known as aldonic acids. D-glucose on oxidation yields gluconic acid.
2. Oxidation of the terminal alcohol group (–CH_2OH) leads to the production of uronic acid. D-glucose on oxidation yields D-glucuronic acid, D-galactose on oxidation yields D-galacturonic acid and D-mannose on oxidation yields D-mannuronic acid.

One of the most important sugar acid is ascorbic acid or vitamin C, which undergoes oxidation to dehydroascorbic acid. Lack of vitamin C in the diet of humans results in scurvy.

3. Oxidation by strong oxidising agents like nitric acid gives aldaric acids, also known as saccharic acid. Both the groups (aldehyde and alcohol) are oxidised. Galactose on oxidation with concentrated nitric acid yields insoluble crystals of mucic acid. This test is used for the identification of galactose.

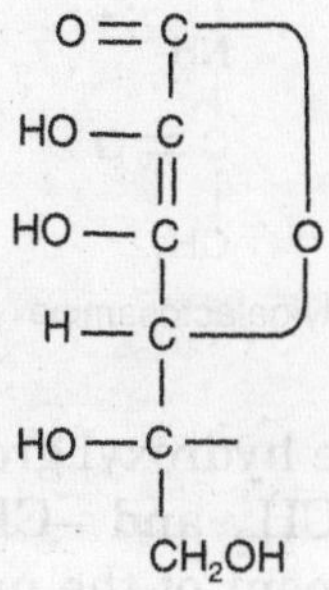

L-Ascorbic acid

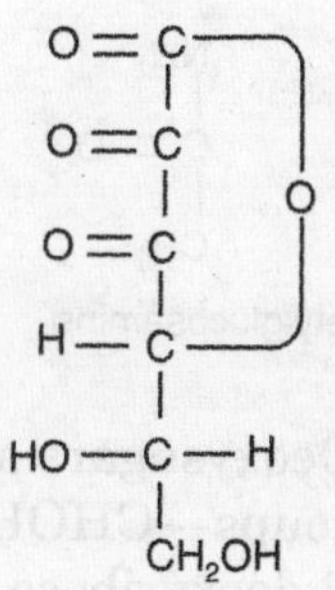

Dehydroascorbic acid

Amino Sugars: When one or more hydroxyl groups of the monosaccharides are replaced by amino groups, they form amino sugars. The most common amino sugar is the D-glucosamine and D-galactosamine and D-mannosamine. They are present as constituent of heteropolysaccharides.

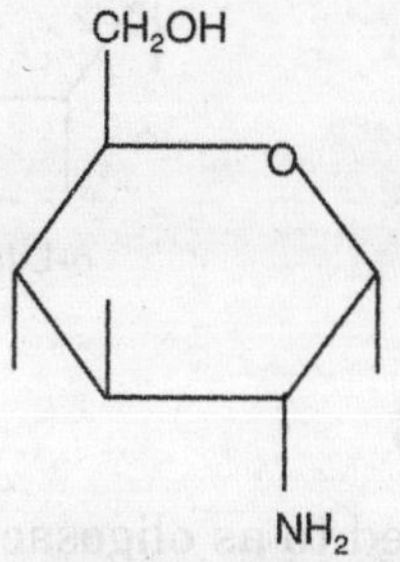

D-glucosamine

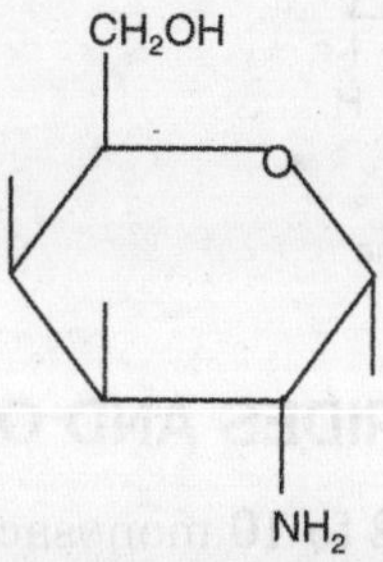

D-galactosamine

The amino groups of such amino sugars are usually acetylated. N-acetylglucosamine and N-acetylgalactosamine are important constituents of chitin and heparin. N-acetylneuraminic (NANA) is a derivative of N-acetylmannose and pyruvic acid. It is an important constituent of glycoproteins and glycolipids. N-acetylmuramic acid is a derivative of N-acetylglucosamine and lactic acid. It is an important constituent of bacterial cell walls.

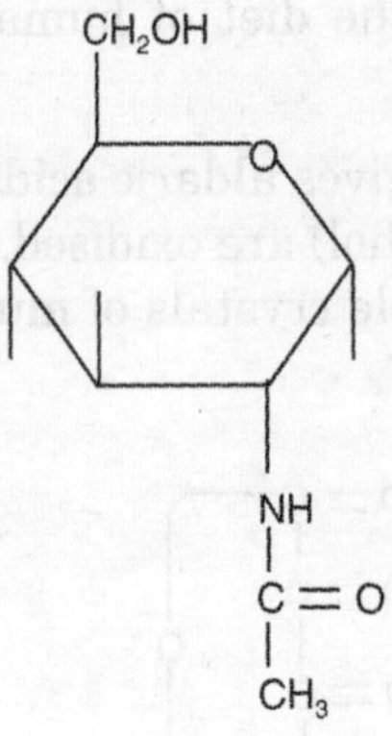

N-acetylglucosamine 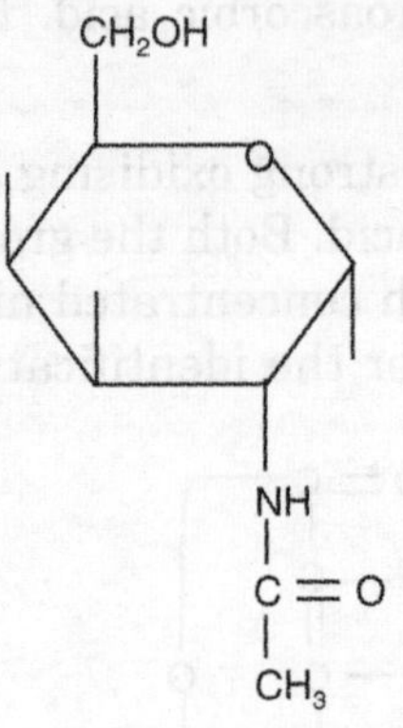 N-acetylgalactosamine

Deoxy Sugars: Deoxysugars are those in which one of the hydroxyl group is replaced by hydrogen (the groups –CHOH and $-CH_2OH$ become $-CH_2$ and $-CH_3$). The most common example is 2-deoxyribose, which is the major component of the nucleic acids. α-L-rhamnose (6-deoxy-α-L-mannose) and α-L-fucose (6-deoxy-α-L-galactose) are the important deoxysugars that occur in polysaccharides.

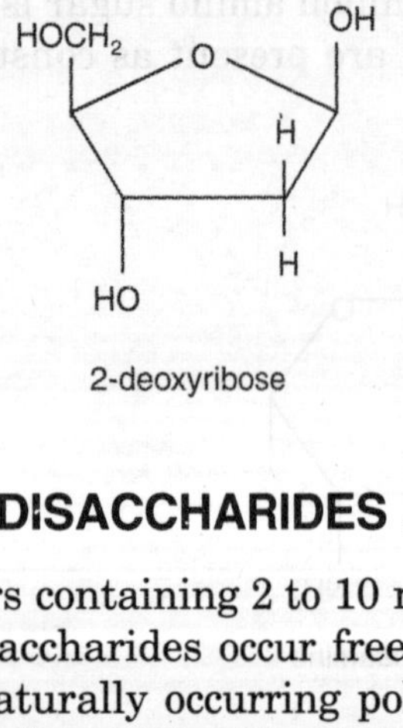

2-deoxyribose 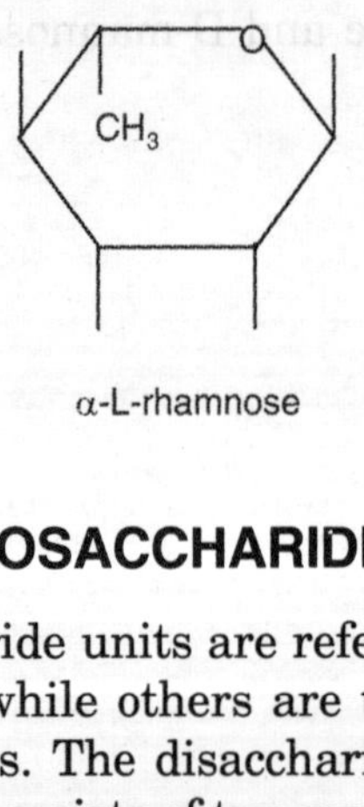 α-L-rhamnose 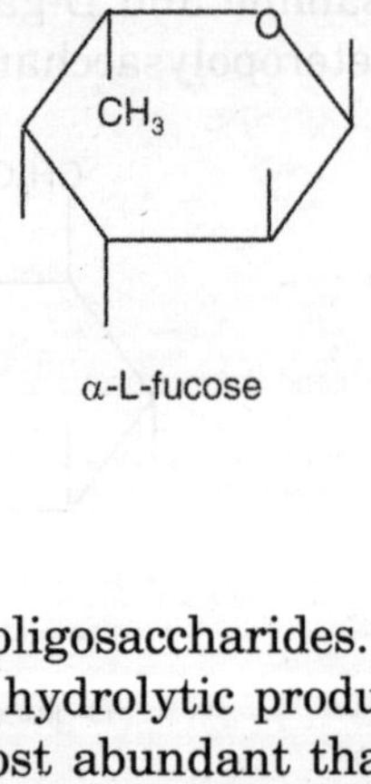 α-L-fucose

4.13 DISACCHARIDES AND OLIGOSACCHARIDES

Sugars containing 2 to 10 monosaccharide units are referred to as oligosaccharides. Some oligosaccharides occur free in nature while others are the partial hydrolytic products of the naturally occurring polysaccharides. The disaccharides are most abundant than the other oligosaccharides. A disaccharide consists of two monosaccharide units (either similar or dissimilar) held together by a glycosidic bond. They are crystalline, water soluble and sweet to taste. The disaccharides are of two types — reducing disaccharides with free aldehyde or keto group (maltose, isomaltose, lactose, cellobiose) and non-reducing disaccharides with no free aldehyde or keto group (sucrose or trehalose).

Maltose: Maltose or malt sugar is not found in free form in the body. It is produced during the course of digestion of starch by the pancreatic amylase enzyme. It consists of two glucose units linked by α (1 → 4) glycosidic bond. It can be hydrolysed by dilute acid or the enzyme maltase to form two molecules of α–D-glucose. Isomaltose, is another disaccharide obtained during the hydrolysis of starch (amylopectin). It is similar to maltose, except that it has an α (1 → 4) glycosidic linkage.

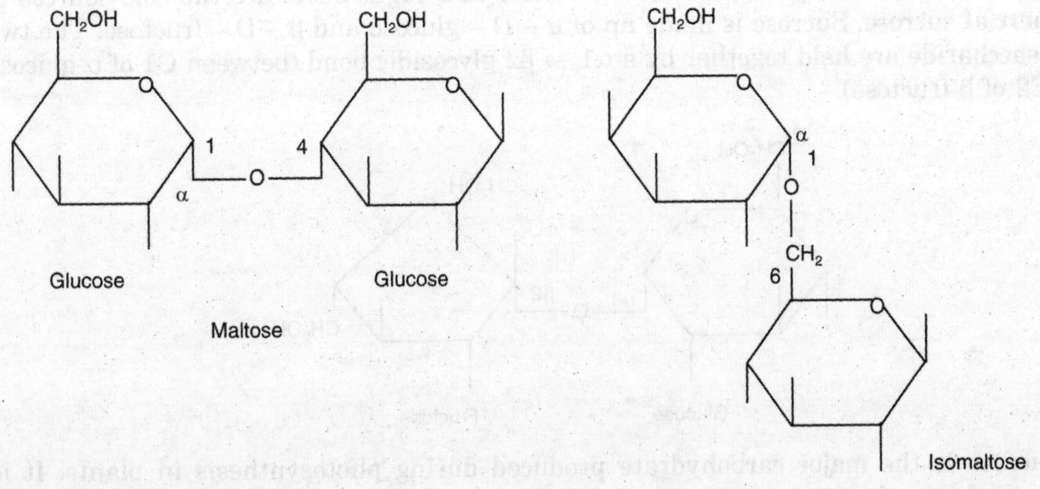

Glucose Glucose

Maltose

Isomaltose

Cellobiose: Cellobiose is the partial hydrolytic product of cellulose. It is identical to maltose except that the linkage is β (1 → 4) glycosidic linkage.

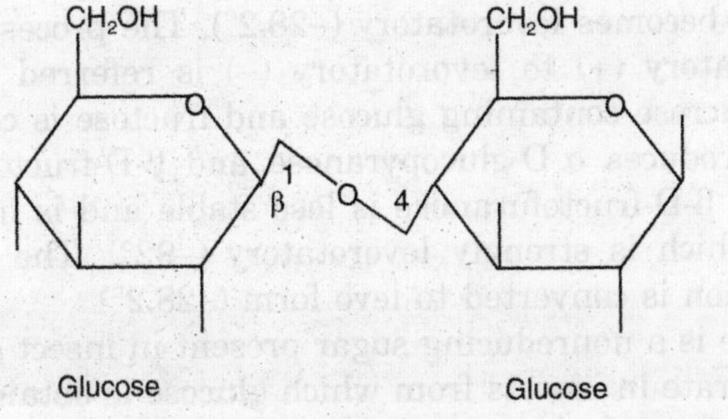

Glucose Glucose

Lactose: Lactose is a disaccharide found in milk. It consists of β–D–galactose and β–D–glucose linked by β (1 → 4) glycosidic bond. Lactose of milk is the most important carbohydrate in the nutrition of young mammals. It is hydrolysed by the intestinal enzyme lactase to glucose and galactose. This enzyme is abundant in the early years of life and gradually disappears with age in certain individuals, causing lactose intolerance.

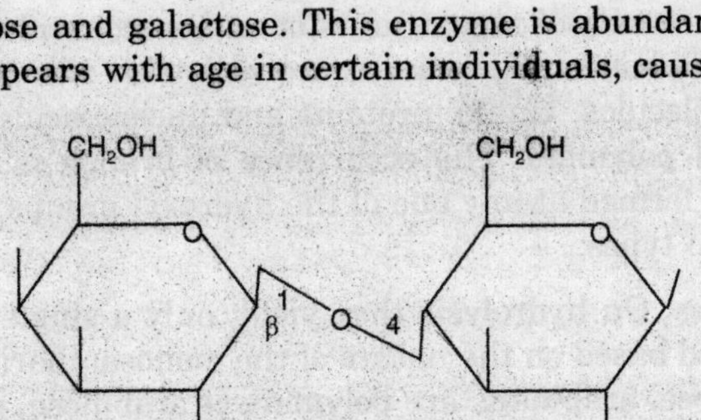

Galactose Glucose

Sucrose: Sucrose is a nonreducing disaccharide of commercial importance and is widely distributed in higher plants. Sugar cane and sugar beets are the sole sources of commercial sucrose. Sucrose is made up of α – D – glucose and β – D – fructose. The two monosaccharide are held together by a $\alpha 1 \rightarrow \beta 2$ glycosidic bond (between C1 of α-glucose and C2 of β-fructose).

CH_2OH CH_2OH O O $\alpha 1$ O $\beta 2$ CH_2OH

Glucose Fructose

Sucrose is the major carbohydrate produced during photosynthesis in plants. It is transported into the storage organs of plants, such as roots, tubers and seeds. It is an important source of dietary carbohydrate. It is sweeter than most other common sugars (except fructose) namely glucose, lactose, and maltose.

Inversion of Sucrose: Sucrose, as such is dextrorotatory (+66.5°). However, when sucrose is hydrolysed, it becomes levorotatory (–28.2°). The process of change in optical rotation from dextrorotatory (+) to levorotatory (–) is referred to as inversion. The hydrolysed mixture of sucrose containing glucose and fructose is called an invert sugar. Hydrolysis of sucrose produces α-D-glucopyranose and β-D-fructofuranose, both being dextrorotatory. However, β-D-fructofuranose is less stable and is immediately converted to β-D-fructopyranose which is strongly levorotatory (–92°). The overall effect is that dextro sucrose on inversion is converted to levo form (–28.2°).

Trehalose: Trehalose is a nonreducing sugar present in insect and fungi. It is essentially a storage carbohydrate in insects from which glucose is obtained. It consists of two glucose units linked by $\alpha 1 \rightarrow \alpha 1$ glycosidic bond.

4.14 POLYSACCHARIDES

Polysaccharides (glycans) consist of repeat units of monosaccharides or their derivatives held together by glycosidic bonds. They are concerned with two important functions – structural and storage molecules. Unlike proteins and nucleic acids, polysaccharides are either linear or branched polymers. The occurrence of branch is due to the fact that glycosidic linkages can be formed at any one of the hydroxyl groups of a monosaccharide. Polysaccharides are of two types:

1. Homopolysaccharides: On hydrolysis they yield only a single type of monosaccharide. They are named based on the nature of the monosaccharide unit. Thus, xylans are polymers of xylose, arabinans are polymers of arabinose, glucans are polymers of glucose, mannans are polymers of mannose, etc.
2. Heteropolysaccharides: On hydrolysis they yield a mixture of a few monosaccharides or their derivatives.

4.14.1 Homopolysaccharides

Starch

Starch, the storage polysaccharide of higher plants consists of polysaccharide granules obtained from the grain of maize, rice, wheat belonging to the family graminae or the tubers of potato belonging to the family solanaceae. It consists of D-glucose units held together by α-glycosidic bonds. Starch is the most important dietary source of higher animals, including man.

Starch consists of two polysaccharides – water-soluble amylose (15 – 20%) and a water insoluble amylopectin (80–85%). Amylose is a linear unbranched chain consisting 200 to 1000 glucose units joined by α (1 → 4) glycosidic bond. Amylopectin consists of a linear chain of α (1 → 4) linked subunits of glucose with α (1 → 6) linked branches. On an average one branch is found for every 25 glucose residues, and each branch contains 20–30 glucose units.

Starches are hydrolysed by pancreatic or salivary amylase to form maltose, glucose and dextrins. Dextrins are the breakdown products of starch and represent the limit of attack by amylases. A debranching enzyme, glucoamylase can hydrolyse α (1 → 6) glycosidic bonds at the branch point. The combined action of amylase and glucoamylase completely degrades amylopectin to maltose and glucose.

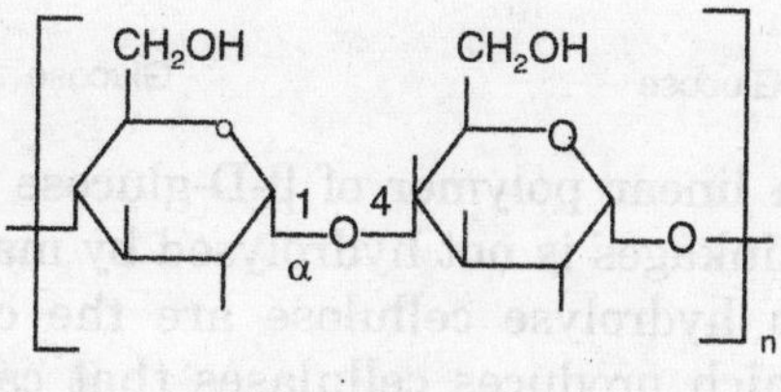

α-amylose

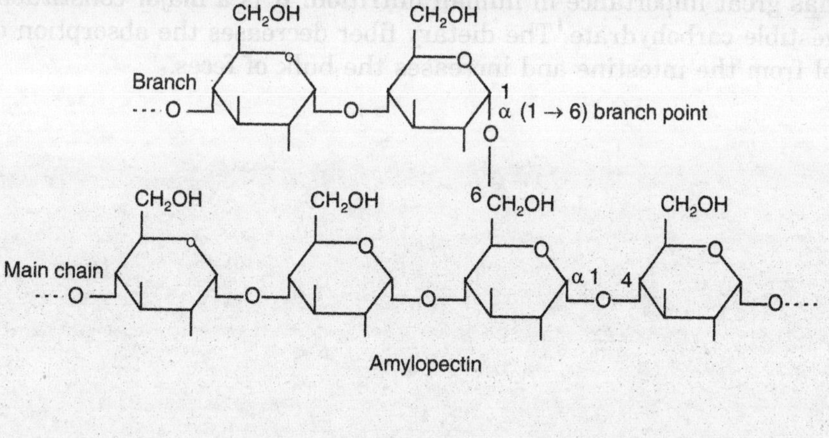

Amylopectin

Glycogen

Glycogen is the main storage polysaccharide of animal cells. It is present in high concentration in liver, followed by muscle, brain, etc. Glycogen is also present in yeast and fungi, that do not contain chlorophyll. Similar to amylopectin of starch, glycogen consists of a linear chain of α (1 → 4) linked subunits of glucose with α (1 → 6) linked branches.

Glycogen is more extensively branched than amylopectin of starch and on an average one branch is found for every 3 – 5 glucose residue. Glycogen is more compact than amylopectin of starch. The molecular weight of glycogen is very high, about 4 to 14 million and corresponds to about 25,000 to 85,000 glucose residues.

Cellulose

Cellulose occurs exclusively in plants and is the most abundant organic substance accounting for about 50% of all carbon. It is a predominant constituent of plant cell wall which serve as structural element, and provides the tensile strength to the wall. Cellulose is present in an almost pure form in cotton fibers (95%), and to a lesser extent in flax (80%), jute (60–70%) and wood (40–50%).

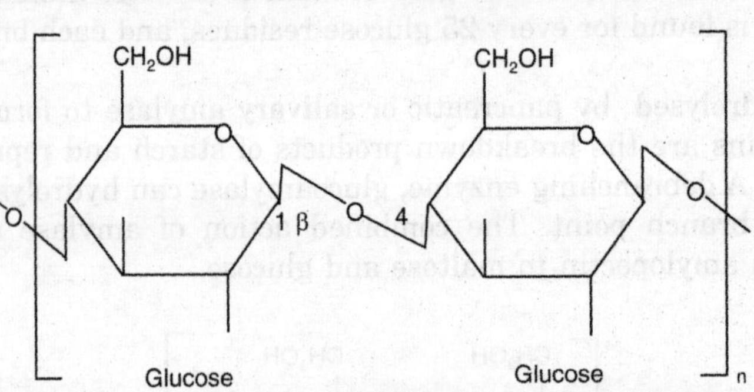

Cellulose is composed of a linear polymer of β-D-glucose units linked by β (1 → 4) glycosidic linkages. β (1 → 4) linkages is not hydrolysed by mammalian enzyme systems. The only vertebrate that can hydrolyse cellulose are the cattles, since they contain microorganisms in the gut which produces cellulases that can hydrolyze cellulose. Hydrolysis of cellulose yields a disaccharide cellobiose, followed by β-D-glucose.

Cellulose has great importance in human nutrition. It is a major constituent of fiber — the non-digestible carbohydrate. The dietary fiber decreases the absorption of glucose and cholesterol from the intestine and increases the bulk of feces.

5

Lipids

5.1 INTRODUCTION

The term lipid is derived from the greek word, lipos meaning fat. The lipids are a large and diverse group of naturally occurring organic compounds that are related by their solubility in nonpolar organic solvents (*e.g.*, ether, chloroform, acetone and benzene) and general insolubility in water. Lipids extracted from biological materials generally represent a mixture of a variety of classes of substances.

Lipids consist of fats, oils, certain vitamins and hormones and most of the non protein membrane components. Lipids together with the carbohydrates and proteins are important as food for many animals. They are also of great biochemical importance because of their role in cellular structure and their role as the chief storage form of energy. In addition, they are commercially important in the form of soaps, detergents, greases and various oils of the paint industry.

5.2 PROPERTIES

Most of the lipids have common structural features that endow them with important biological properties.

1. They are mainly ionic or polar derivatives of hydrocarbons and belong to the class of compounds called amphiphiles (amphi means "both", phile means "affinity"). Amphiphiles contain polar or ionic groups that have affinity for water, as well as nonpolar hydrocarbon groups that lack affinity for water.
2. The properties of amphiphiles are influenced by the nature of these groups. Thus, certain lipids such as natural fats are only weakly polar and have little affinity for water. They are stored in tissues largely in an anhydrous form and serve as reservoir of energy.
3. Other lipids such as the phosphoglycerides, glycolipids and sphingolipids are more polar and due to their ampliphilic properties form major structural components of the various biological membrane that serve to compartmentalise the cells.

5.3 LIPIDS FUNCTIONS

1. Fats and lipids are important because they serve as energy source, as well as a storage for energy in the form of fat cells.
2. Lipids have a major cellular function as structural components in cell membranes. These membranes in association with carbohydrates and proteins regulate the flow of water, ions and other molecules into and out of the cells.
3. Hormones, steroids, and prostaglandins are chemical messengers between body tissues.
4. Vitamins A, D, E and K are lipid-soluble and regulate critical biological processes. Other lipids aid in vitamin absorption and transportation.
5. Lipids act as a shock absorber to protect vital organs and insulate the body from temperature extremes.

Thus, a study of the structures and properties of the various lipids are important to understand their diverse biological function.

5.4 CLASSIFICATION

There is no single, internationally accepted system of classification for the lipids. However, lipids are conveniently classified as follows:

1. Fatty acids
2. Lipids containing glycerol
   (a) Neutral fats
   (b) Phospholipids
3. Lipids not containing glycerol
   (a) Sphingolipids
   (b) Aliphatic alcohols and waxes
   (c) Prostaglandins, Thromboxanes and Leukotrienes
   (d) Terpenes
   (e) Steroids
4. Lipids combined with other classes of compounds.
   (a) Lipoproteins

5.4.1 Fatty Acids

The common feature of these lipids is that they are all esters of moderate to long chain fatty acids. Acid or base-catalysed hydrolysis yields the component fatty acid (some examples of which are given in the following table) together with the alcohol component of the lipid. These long-chain carboxylic acids are generally referred to by their common names, which in most cases reflect their sources. Natural fatty acids may be saturated or unsaturated, and as the following data indicate, the saturated acids have higher melting points than unsaturated acids of corresponding size. The double bonds in the unsaturated compounds listed in the table are all cis (or Z).

Fatty Acids					
Saturated			Unsaturated		
Formula	Common Name	Melting Point	Formula	Common Name	Melting Point
$CH_3(CH_2)_{10}CO_2H$	lauric acid	45°C	$CH_3(CH_2)_5CH{=}CH(CH_2)_7CO_2H$	palmitoleic acid	0°C
$CH_3(CH_2)_{12}CO_2H$	myristic acid	55°C	$CH_3(CH_2)_7CH{=}CH(CH_2)_7CO_2H$	oleic acid	13°C
$CH_3(CH_2)_{14}CO_2H$	palmitic acid	63°C	$CH_3(CH_2)_4CH{=}CHCH_2CH{=}CH(CH_2)_7CO_2H$	linoleic acid	–5°C
$CH_3(CH_2)_{16}CO_2H$	stearic acid	69°C	$CH_3CH_2CH{=}CHCH_2CH{=}CHCH_2CH{=}CH(CH_2)_7CO_2H$	linolenic acid	–11°C
$CH_3(CH_2)_{18}CO_2H$	arachidic acid	76°C	$CH_3(CH_2)_4(CH{=}CHCH_2)_4\ CH_2)_2CO_2H$	arachidonic acid	–49°C

5.4.1.1 Properties

The unsaturated fatty acids have lower melting points than the saturated fatty acids. The reason for this phenomenon can be found by a careful consideration of molecular geometries. The tetrahedral bond angles on carbon results in a molecular geometry for saturated fatty acids that is relatively linear although with zigzags. This molecular structure allows many fatty acid molecules to be rather closely "stacked" together. As a result, close intermolecular interactions result in relatively high melting points.

On the other hand, the introduction of one or more double bonds in the hydrocarbon chain in unsaturated fatty acids results in one or more "bends" in the molecule. The geometry of the double bond is almost always a cis configuration in natural fatty acids. These molecules do not "stack" very well. The intermolecular interactions are much weaker than saturated molecules. As a result, the melting points are much lower for unsaturated fatty acids.

The trans-double bond isomer of oleic acid, known as elaidic acid, has a linear shape and a melting point of 45° C (32° C higher than its cis isomer). The shapes of stearic and oleic acids are displayed in the models below.

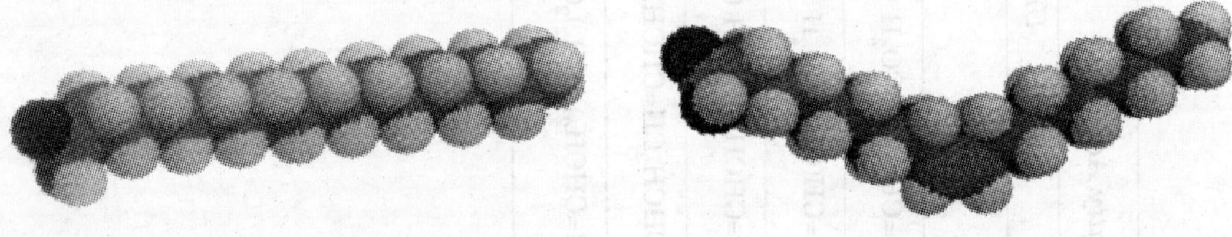

Stearic acid Oleic acid

Two polyunsaturated fatty acids, linoleic and linolenic, are designated "essential" because their absence in the human diet has been associated with health problems, such as scaly skin, stunted growth and increased dehydration. These acids are also precursors to the prostaglandins — a family of physiologically potent lipids present in minute amounts in most body tissues.

In vertebrates, free fatty acids having a free carboxylate group circulate in the blood bound non-covalently to a protein carrier — serum albumin. Fatty acids are also present in blood plasma mostly as carboxylic acid derivatives such as esters or amides, and these are even less soluble in water than are the free fatty acids.

Fatty acids, due to their enhanced acidity, react with bases to form ionic salts. In the case of alkali metal hydroxides the resulting salts have pronounced ionic character and are usually soluble in water.

$$RCO_2H + NaOH \longrightarrow RCO_2^{(-)}Na^{(+)} + H_2O$$

$$RCO_2H + KOH \longrightarrow RCO_2^{(-)}K^{(+)} + H_2O$$

5.4.1.2 Soaps and Detergents

Fatty acids and salts having alkyl chains longer than eight carbons exhibit unusual behaviour in water due to the presence of both hydrophilic (CO_2^-) and hydrophobic (alkyl) regions in the same molecule. Such molecules are termed amphiphilic or amphipathic. Fatty acids made up of ten or more carbon atoms are nearly insoluble in water, and because of their lower density, float on the surface when mixed with water. Unlike paraffin or other alkanes, which tend to puddle on the waters surface, these fatty acids spread evenly over an extended water surface, eventually forming a monomolecular layer in which the polar carboxyl groups are hydrogen bonded at the water interface, and the hydrocarbon chains are aligned together away from the water. This behaviour is illustrated in the diagram below. Substances that accumulate at water surfaces and change the surface properties are called surfactants.

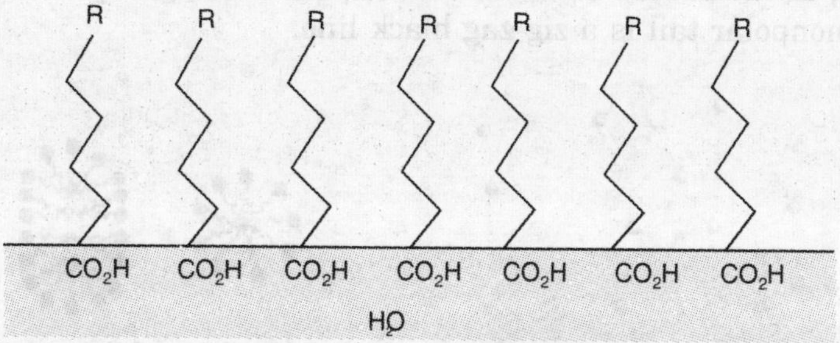

R R R R R R R

CO_2H CO_2H CO_2H CO_2H CO_2H CO_2H CO_2H

HO

Alkali metal salts of fatty acids are more soluble in water than the acids themselves, and the amphiphilic character of these substances also make them strong surfactants. The most common examples of such compounds are soaps and detergents, four of which are shown below. Each of these molecules has a nonplanar hydrocarbon chain, the "tail"

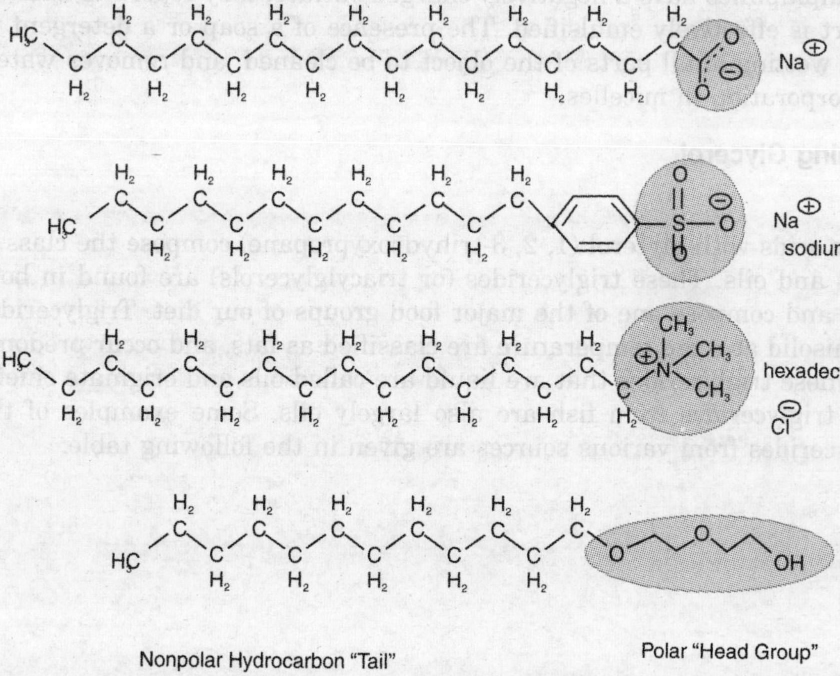

a soap
sodium stearate

an anionic detergent
sodium p-dodecylbenzenesulfonate

a cationic detergent
hexadecyltrimethylammonium chloride

a nonionic detergent
di(ethyl glycol) dodecyl ether

Nonpolar Hydrocarbon "Tail"

Polar "Head Group"

and a polar (often ionic) "head group". The use of such compounds as cleaning agents is facilitated by their surfactant character, which lowers the surface tension of water, allowing it to penetrate and wet a variety of materials.

Very small amounts of these surfactants dissolve in water to give a random dispersion of solute molecules. However, when the concentration is increased, an interesting change occurs. The surfactant molecules reversibly assemble into polymolecular aggregates called micelles. By gathering the hydrophobic chains together in the center of the micelle, disruption of the hydrogen bonded structure of liquid water is minimised, and the polar head groups extend into the surrounding water where they participate in hydrogen bonding. These micelles are often spherical in shape, but may also assume cylindrical and branched forms, as illustrated below. Here the polar head group is designated by a black circle, and the nonpolar tail is a zig-zag black line.

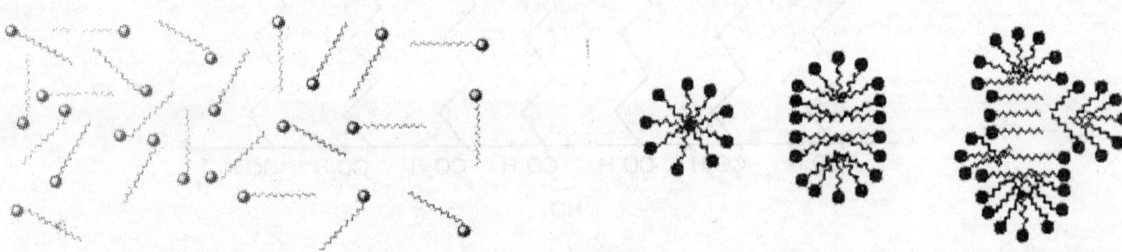

Micelles are able to encapsulate nonpolar substances such as grease within their hydrophobic center, and thus solubilise it so it is removed with the wash water. Since the micelles of anionic amphiphiles have a negatively charged surface, they repel one another and the nonpolar dirt is effectively emulsified. The presence of a soap or a detergent in water facilitates the wetting of all parts of the object to be cleaned, and removes water-insoluble dirt by incorporation in micelles.

5.4.2 Lipids Containing Glycerol

5.4.2.1 Neutral Fats

The triesters of fatty acids with glycerol (1, 2, 3-trihydroxypropane) compose the class of lipids known as fats and oils. These triglycerides (or triacylglycerols) are found in both plants and animals, and compose one of the major food groups of our diet. Triglycerides that are solid or semisolid at room temperature are classified as fats, and occur predominantly in animals. Those triglycerides that are liquid are called oils and originate chiefly in plants, although triglycerides from fish are also largely oils. Some examples of the composition of triglycerides from various sources are given in the following table:

Source	Saturated Acids (%) C_{10} & less	C_{12} lauric	C_{14} myristic	C_{16} palmitic	C_{18} stearic	Unsaturated Acids (%) C_{18} oleic	C_{18} linoleic	C_{18} unsaturated
				Animal Fats				
butter	15	2	11	30	9	27	4	1
lard	–	–	1	27	15	48	6	2
human fat	–	1	3	25	8	46	10	3
herring oil	–	–	7	12	1	2	20	52
				Plant Oils				
Coconut	–	50	18	8	2	6	1	–
Corn	–	–	1	10	3	50	34	–
Olive	–	–	–	7	2	85	5	–
Palm	–	–	2	41	5	43	7	–
Peanut	–	–	–	8	3	56	26	7
Safflower	–	–	–	3	3	19	76	–

Fats have a predominance of saturated fatty acids, and oils are composed largely of unsaturated fatty acids. Thus, the melting points of triglycerides reflect their composition. Natural mixed triglycerides have somewhat lower melting points, the melting point of lard being near 30°C, whereas olive oil melts near –6°C. Since fats are valued over oils by some Northern European and North American populations, vegetable oils are extensively converted to solid triglycerides by partial hydrogenation of their unsaturated components. Some of the remaining double bonds are isomerised (to trans) in this operation. These saturated and trans-fatty acid glycerides in the diet have been linked to long-term health issues such as atherosclerosis.

$$
\begin{array}{l}
H_2C-OCO(CH_2)_{10}CH_3 \\
HC-OCO(CH_2)_{10}CH_3 \\
H_2C-OCO(CH_2)_{10}CH_3
\end{array}
\qquad
\begin{array}{l}
H_2C-OCO(CH_3)_{16}CH_3 \\
HC-OCO(CH_3)_{16}CH_3 \\
H_2C-OCO(CH_3)_{16}CH_3
\end{array}
\qquad
\begin{array}{l}
H_2C-OCO(CH_2)_7CH\overset{cis}{=}CH\,(CH_2)_7CH_3 \\
HC-OCO(CH_2)_7CH\overset{cis}{=}CH\,(CH_2)_7CH_3 \\
H_2C-OCO(CH_2)_7CH\overset{cis}{=}CH\,(CH_2)_7CH_3
\end{array}
$$

trilaurin mp 45 °C — tristearin mp 71 °C — triolein mp –4 °C

Triglycerides having three identical acyl chains, such as tristearin and triolein (above), are called "simple" while those composed of different acyl chains are called "mixed". If the acyl chains at the end hydroxyl groups (1 & 3) of glycerol are different, the center carbon becomes a chiral center and enantiomeric configurations must be recognised.

In most eukaryotic cells, triacylglycerides form a separate phase of microscopic, oily droplets in the aqueous cytosol and serve as a depots of metabolic fuel.

In invertebrates, specialised cells called adipocytes or fat cells store large amounts of triacylglycerides as droplets that nearly fill the cell.

Triacylglycerides are also stored as oils in the seeds of many types of plants and they provide energy and biosynthetic precursors during seed germination.

Humans have fat tissues, which is primarily of adipocytes under the skin and in the abdominal cavity. Moderately obese people with 15-20 kg of triacylglycerides deposited in their adipocytes could meet their energy needs for months by drawing on their fat stores. In contrast, the human body can store less than a days energy supply in the form of glycogen.

In some animals, triacylglycerides stored under the skin not only serve as energy stores, but as insulation against low temperature. Seals, Penguins and other warm-blooded polar animals are amply padded with triacylglycerides.

In hibernating animals (bears), the huge fat reserves accumulated before hibernation serve the dual purpose of insulation and energy storage. Fat bears carry on β-oxidation even in their sleep.

5.4.2.1.1 Hydrolysis: Hydrolysis of the neutral fat yields three molecules of fatty acids and one molecule of glycerol. The reaction proceeds slowly in boiling water, but is greatly accelerated by catalytic amounts of acid or base. The enzymes that catalyse this reaction in animals and plants are esterases or more specifically lipases.

$$\begin{array}{l} CH_2-\overset{O}{\overset{\|}{C}}-O-R' \\ | \\ CH-\overset{O}{\overset{\|}{C}}-O-R'' \\ | \\ CH_2-\overset{O}{\overset{\|}{C}}-O-R''' \end{array} + 3H_2O \rightleftharpoons \begin{array}{c} R'-COOH \\ + \\ R''-COOH \\ + \\ R'''-COOH \end{array} + \begin{array}{c} CH_2OH \\ | \\ CH_2OH \\ | \\ CH_2OH \end{array}$$

5.4.2.1.2 Saponification: Hydrolysis of neutral fats by base is termed saponification. The carboxylate ions formed in the presence of a cation (Na^+ or K^+) becomes soap. The reaction is irreversible. The saponifiable fraction is defined as that portion of total lipids which after treatment with hot alkali is soluble in water and insoluble in ether. Neutral fats are thus said to be saponifiable.

$$\begin{array}{l} CH_2-\overset{O}{\overset{\|}{C}}-O-R \\ | \\ CH-\overset{O}{\overset{\|}{C}}-O-R \\ | \\ CH_2-\overset{O}{\overset{\|}{C}}-O-R \end{array} + 3NaOH \longrightarrow R-\overset{O}{\overset{\|}{C}}-O^-Na^+ + \begin{array}{c} CH_2OH \\ | \\ CH_2OH \\ | \\ CH_2OH \end{array}$$

Saponification is a method of characterising an oil or fat. It indicates the size of the component fatty acids. Saponification number is the weight in milligrams of KOH required to completely saponify one gram of an oil or fat.

5.4.2.1.3 Halogenation: The unsaturated fatty acids present in an oil or fat can combine with halogens like bromine and iodine. The addition of halogen is used to determine the degree of unsaturation (Iodine Number).

Iodine number is the weight in grams of iodine that can be taken up by 100 gm fat or oil. Higher the iodine number, greater will be the unsaturation. Oils have higher iodine number than fats.

Oil or Fat	*Saponification number*	*Iodine number*
Coconut oil	253-262	6.2–10
Butter fat	210-230	26–38
Groundnut oil	186-194	88–98
Castor oil	175-182	84–88
Sunflower oil	188-193	129–136
Palm oil	186-194	49–59

5.4.2.1.4 Oxidation: Oil and fats in contact with air, become rancid — develop unpleasant taste and odour. The double bonds in the unsaturated fatty acids are oxidised to yield short chain aldehydes and keto acids which have bad odour. This type of rancidity is called oxidative rancidity. This can be prevented by the addition of anti-oxidants like, tocopherols (Vitamin E), citric acid, ascorbic acid (Vitamin C).

Rancidity can also occur due to partial hydrolysis of the ester linkage giving rise to short chain fatty acids. This type of rancidity is called as hydrolytic rancidity.

5.4.2.2 Phospholipids

Phospholipids are similar to the triglycerides with a couple of exceptions. Phosphoglycerides are esters of only two fatty acids — phosphoric acid and a trifunctional alcohol — glycerol (IUPAC name is 1,2,3-propantriol). The fatty acids are attached to the glycerol at the 1 and 2 positions on glycerol through ester bonds. There may be a variety of fatty acids, both saturated and unsaturated, in the phospholipids.

The third oxygen on glycerol is bonded to phosphoric acid through a phosphate ester bond (oxygen-phosphorus double bond oxygen). In addition, there is usually a complex amino alcohol also attached to the phosphate through a second phosphate ester bond. The complex amino alcohols include choline, ethanolamine, and the aminoacid–serine.

Phospholipid Components

HO, OH, OH — glycerol

Fatty Acids saturated & unsaturated

HO, NH_2 — ethanolamine

O, HO, C $(CH_2)_{12}CH_3$ — fatty acid

O, OH, P, HO, OH — phosphoric acid

HO, $\overset{\oplus}{N}(CH_3)_3$ $\overset{\ominus}{O}H$ — choline hydroxide

The properties of a phospholipid are characterised by the properties of the fatty acid chain and the phosphate/amino alcohol. The long hydrocarbon chains of the fatty acids are non-polar. The phosphate group has a negatively charged oxygen and a positively charged nitrogen to make this group ionic. In addition, there are other oxygen of the ester groups, which make on whole end of the molecule strongly ionic and polar.

Phospholipids are major components in the lipid bilayers of cell membranes. There are two common phospholipids:

1. Lecithin contains the amino alcohol — choline
2. Cephalins contain the amino alcohols serine or ethanolamine

1. Lecithin

Lecithin is the most common phospholipid. It is found in egg yolks, wheat germ, and soybeans. Lecithin is extracted from soy beans for use as an emulsifying agent in foods. Lecithin is an emulsifier because it has both polar and non-polar properties, which enable it to cause the mixing of other fats and oils with water components. Lecithin is also a major component in the lipid bilayers of cell membranes. Lecithin contains choline joined to the phosphate by an ester linkage. The nitrogen has a positive charge, just as in the ammonium ion.

2. Cephalins

Cephalins are phosphoglycerides that contain ethanolamine or the amino acid serine attached to the phosphate group through phosphate ester bonds. A variety of fatty acids make up the rest of the molecule. Cephalins are found in most cell membranes, particularly in brain tissues. They are also important in the blood clotting process as they are found in blood platelets.

Cephalin Phospholipid

$H-C-O-C(=O)-(CH_2)_{16}CH_3$
$H-C-O-C(=O)-(CH_2)_{16}CH_3$ — Non-Polar
$H-C-O-P(=O)(O^-)-O-CH_2CH_2-\overset{+}{N}H_3$ — ethanolamine — Polar

Lecithin Phospholipid

$H-C-O-C(=O)-(CH_2)_{16}CH_3$
$H-C-O-C(=O)-(CH_2)_{16}CH_3$ — Non-Polar
$H-C-O-P(=O)(O^-)-O-CH_2CH_2-\overset{+}{N}(CH_3)_3$ — Choline — Polar

5.4.2.2.1 Properties of Phospholipids: As ionic amphiphiles, phospholipids aggregate or self-assemble when mixed with water, but in a different manner than the soaps and detergents. Because of the two pendant alkyl chains present in phospholipids and the unusual mixed charges in their head groups, micelle formation is unfavourable relative to a bilayer structure. If a phospholipid is smeared over a small hole in a thin piece of plastic immersed in water, a stable planar bilayer of phospholipid molecules is created at the hole. The polar head groups on the faces of the bilayer contact water, and the hydrophobic alkyl chains form a nonpolar interior. The phospholipid molecules can move about in their half the bilayer, but there is a significant energy barrier preventing migration to the other side of the bilayer. This bilayer membrane structure is also found in aggregate structures called liposomes. Liposomes are microscopic vesicles consisting of an aqueous core enclosed in one or more phospholipid layers. They are formed when phospholipids are vigorously mixed with water. Unlike micelles, liposomes have both aqueous interiors and exteriors.

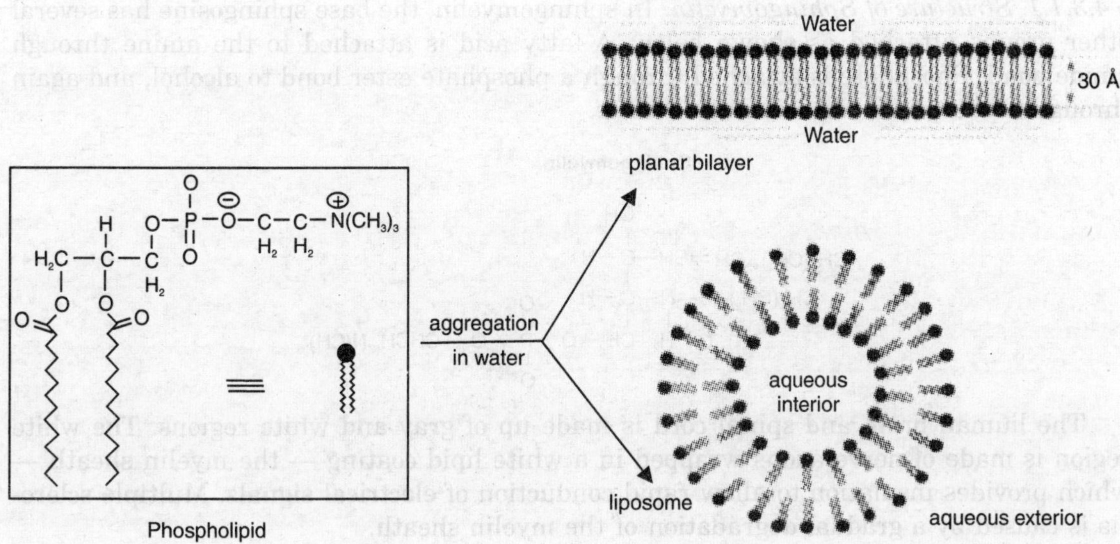

A cell may be considered a very complex liposome. The bilayer membrane that separates the interior of a cell from the surrounding fluids is largely composed of phospholipids, but it incorporates many other components, such as cholesterol, that contribute to its structural integrity. Protein channels that permit the transport of various kinds of chemical species in and out of the cell are also important components of cell membranes. The interior of a cell contains a variety of structures (organelles) that conduct chemical reactions vital to the cells existence.

5.4.3 Lipids Not Containing Glycerol

5.4.3.1 Sphingolipids

Sphingolipids are named after the spinx in Greek mythology, part woman and part lion, who devoured all who could not answer her riddles. Sphingolipids appeared to Johann Thudichum in 1874 as part of the dangerous riddle of the brain.

Sphingolipids are a second type of lipid found in cell membranes, particularly nerve cells and brain tissues. They do not contain glycerol, but retain the two alcohols with the middle position occupied by an amine. They contain either sphingosine (D-4-sphingenine) or Dihydrosphingosine (D-sphinganine) as aliphatic base (1, 3-dihydroxy-2-aminoalcohol). Sphingolipids containing 18-carbon atoms are most abundant. However, other homologous such as 16-C, 18-C and 20-C are also found among naturally occurring sphingolipids. The double bond has the transconfiguration and the asymmetric centres C-2 and C-3 are of the D-configuration. Sphingosine has three parts — a three carbon chain with two alcohols and amine attached and a long hydrocarbon chain.

Sphingosine

$$CH_3(CH_2)_{12}CH{=}CH{-}CH(OH){-}CH(NH_2){-}CH_2OH$$

5.4.3.1.1 Structure of Sphingomyelin: In sphingomyelin, the base sphingosine has several other groups attached as shown below. A fatty acid is attached to the amine through amide bond. Phosphate is attached through a phosphate ester bond to alcohol, and again through a phosphate ester bond to choline.

Sphingomyelin

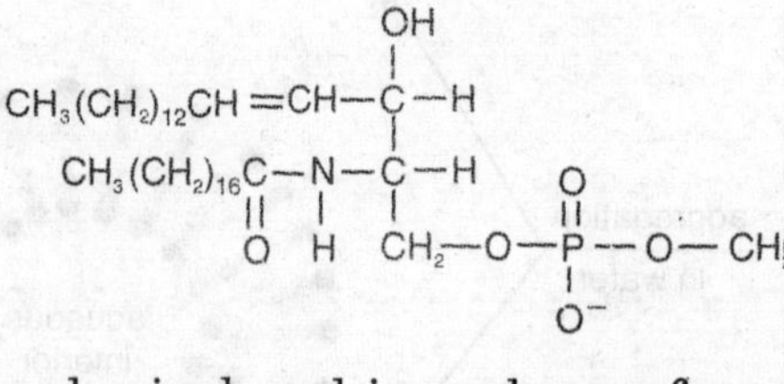

$$\begin{array}{l} \qquad\qquad\qquad\qquad\quad OH \\ CH_3(CH_2)_{12}CH{=}CH{-}C{-}H \\ CH_3(CH_2)_{16}C{-}N{-}C{-}H \qquad O \\ \qquad\qquad\;\; \| \quad | \quad\;\; | \qquad\quad \| \qquad\quad + \\ \qquad\qquad\;\; O \;\; H \;\; CH_2{-}O{-}P{-}O{-}CH_2CH_2N(CH_3)_3 \\ \qquad\qquad\qquad\qquad\qquad\qquad\;\; O^- \end{array}$$

The human brain and spinal cord is made up of gray and white regions. The white region is made of nerve axons wrapped in a white lipid coating — the myelin sheath — which provides insulation to allow rapid conduction of electrical signals. Multiple sclerosis is caused by a gradual degradation of the myelin sheath.

Sphingomyelins are located throughout the body in nerve cell membranes. They make up about 25% of the lipids in the myelin sheath that surrounds and insulates cells of the central nervous system. Niemann-Pick disease is caused by a deficiency of an enzyme that breaks down excessive sphingomyelin, which then builds up on the liver, spleen, brain and bone marrow. An affected child usually dies within several years.

5.4.3.1.2 Glycolipids and Cerebrosides: Glycolipids are complex lipids that contain carbohydrates. Cerebrosides are an example which contain the sphingosine backbone attached to a fatty acid and a carbohydrate. The carbohydrates are most often glucose or galactose. Those that contain several carbohydrates are called gangliosides. Glucocerebroside has the specific function in the cell membranes of macrophages (cells that protect the body by destroying foreign microorganisms). Galactocerebroside is found almost exclusively in the membranes of brain cells.

There are several genetic diseases resulting from the absence of specific enzymes which breakdown the glycolipids. Tay-Sachs, which mainly affect Jewish children, results in a build up of gangliosides and result in death in several years. Gaucher's disease results in the excessive build up of glucocerebroside resulting in severe anemia and enlarged liver and spleen.

Glucocerebroside

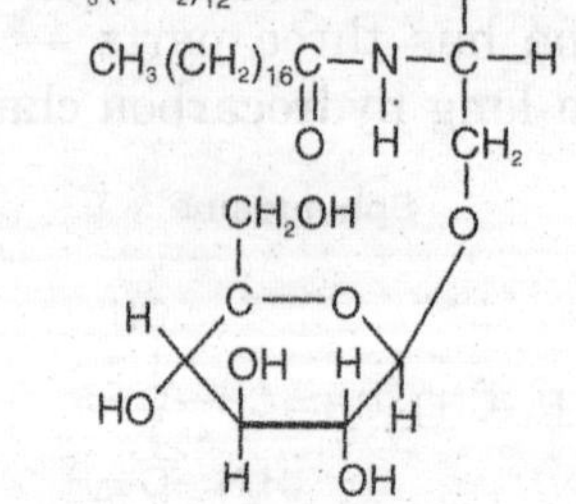

$$\begin{array}{l} \qquad\qquad\qquad\qquad\quad OH \\ CH_3(CH_2)_{12}CH{=}CH{-}C{-}H \\ CH_3(CH_2)_{16}C{-}N{-}C{-}H \\ \qquad\qquad\;\; O \;\; H \;\; CH_2 \\ \qquad\quad CH_2OH \qquad O \\ \qquad\quad C{-}O \\ H \qquad OH \;\; H \\ HO \qquad\qquad\quad H \\ \qquad\quad H \qquad OH \end{array}$$

5.4.3.2 Aliphatic Alcohols and Waxes

Significant quantities of aliphatic alcohols (containing 16–30 carbon atoms) have been recovered from certain lipid sources. Wool fat, obtained from the wool of sheeps, contains aliphatic alcohols such as cetyl alcohol (C-16) and ceryl alcohol (C-26).

Cetyl alcohol	$CH_3(CH_2)_{14}CH_2OH$
Ceryl alcohol	$CH_3(CH_2)_{24}CH_2OH$

A wax is a simple lipid which is an ester of a long-chain alcohol and a long chain fatty acid. The alcohol may contain from 12-32 carbon atoms.

Waxes are widely distributed in nature and are the chief storage form of metabolic fuel in free-floating marine microorganisms. They are also found in nature as coatings on leaves and stems. The wax prevents the plant from losing excessive amounts of water and protect them from dehydration and small predators. Carnuba wax is valued for its toughness and water resistance and is found on the leaves of Brazilian palm trees and is used in floor and automobile waxes. Lanolin coats lambs wool. Beeswax is secreted by bees to make cells for honey and eggs. Spermaceti wax is found in the head cavities and blubber of the sperm whale. The feathers of birds and the fur of some animals have similar coatings which serve as a water repellent. Certain skin glands of vertebrates secrete waxes to protect hair and skin, and keep it pliable, lubricated and water proof. Many of the waxes mentioned are used in ointments, hand creams, and cosmetics.

Paraffin wax, used in some candles, is not based upon the ester functional group, but is a mixture of high molecular weight alkanes. Ear wax is a mixture of phospholipids and esters of cholesterol.

Waxes are esters of fatty acids with long chain monohydric alcohols (one hydroxyl group). Natural waxes are often mixtures of such esters, and may also contain hydrocarbons.

spermaceti wax	$CH_3(CH_2)_{14}CO_2\text{-}(CH_2)_{15}CH_3$
bees wax	$CH_3(CH_2)_{24}CO_2\text{-}(CH_2)_{29}CH_3$
carnuba wax	$CH_3(CH_2)_{30}CO_2\text{-}(CH_2)_{33}CH_3$

5.4.3.3 Prostaglandins, Thromboxanes and Leukotrienes

All naturally occurring prostaglandins are present nearly in all mammalian tissues, but in very low concentrations. They belong to a class of organic compounds which are derived from C-20 unsaturated fatty acid, arachidonic acid (5, 8, 11, 14-eicosatetraenoic acid), by cyclization. Prostaglandins and their related compounds — the thromboxanes and leukotrienes are collectively known as eicosanoids, since all of them contain 20 carbon atoms (in greek, eikus means twenty). The name prostaglandin was first suggested by Swedish physiologist V.S. von Euler, since it was first discovered in the secretions of prostrate gland. The parent compound of prostaglandins is prostanoic acid, which is a C-20 unsaturated fatty acid with a cyclopentane ring between C-8 and C-12. The side chain at C-8 has α-configuration, while the side chain at C-12 has β-configuration. The hydroxyl group at C-11 and C-15 has α-configuration and the double bond between C-13 and C-14 is of trans configuration.

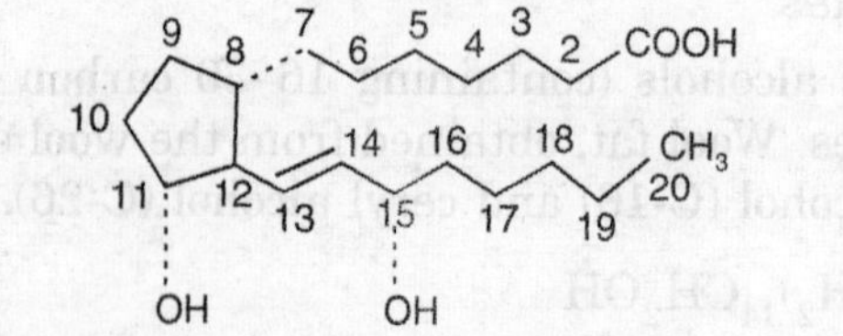

Prostanoic acid

There are mainly two series of prostaglandins — prostaglandin E, and prostaglandin F. They differ only in the presence of a ketone in the E series and an hydroxyl group in the F series at position C-9.

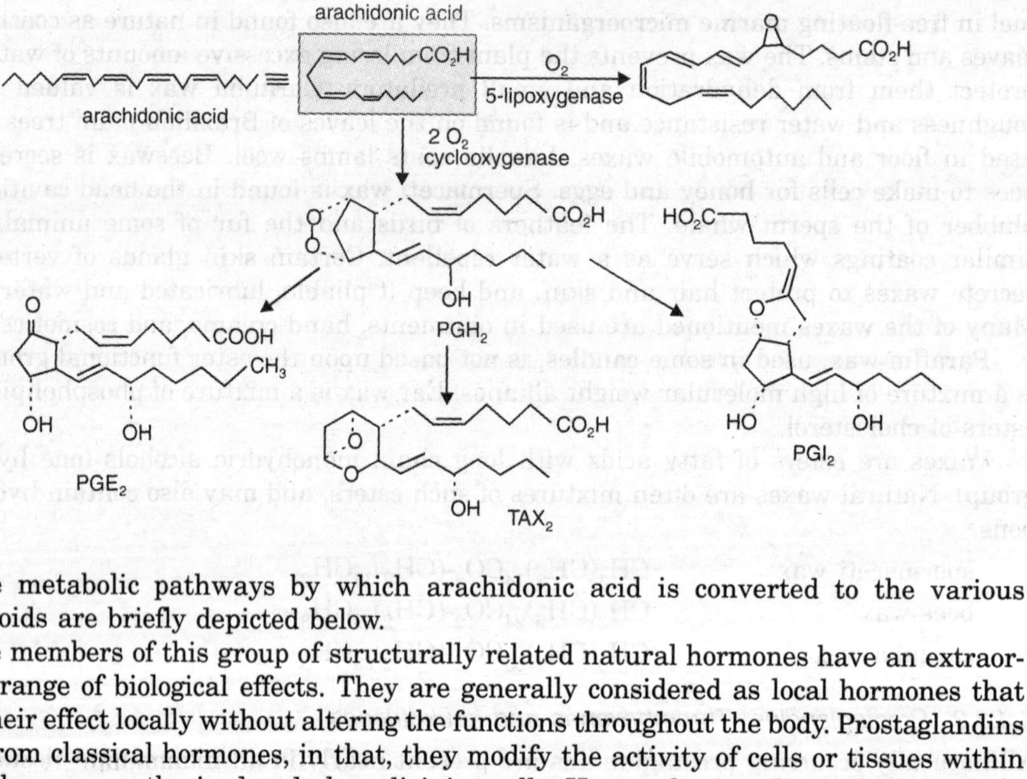

arachidonic acid

O_2 5-lipoxygenase

2 O_2 cyclooxygenase

PGH_2 PGE_2 PGI_2 TAX_2

The metabolic pathways by which arachidonic acid is converted to the various eicosanoids are briefly depicted below.

The members of this group of structurally related natural hormones have an extraordinary range of biological effects. They are generally considered as local hormones that exert their effect locally without altering the functions throughout the body. Prostaglandins differ from classical hormones, in that, they modify the activity of cells or tissues within which they are synthesised and also adjoining cells. Hence, they are known as paracrine hormones.

There are a variety of physiological effects including:

1. Activation of the inflammatory response, production of pain, and fever.
2. Blood clots form when a blood vessel is damaged. A type of prostaglandin called thromboxane stimulates vascular constriction and aggregation of platelets. Conversely, PGI2, is produced to have the opposite effect on the walls of blood vessels where clots should not be forming.
3. Certain prostaglandins are involved with the induction of labor and other reproductive processes. PGE2 causes uterine contractions and has been used to induce labor.

4. Prostaglandins are involved in several other organs, such as the gastrointestinal tract (inhibit acid synthesis and increase secretion of protective mucus), increase blood flow in kidneys, and leukotriens promote constriction of bronchi associated with asthma.

5.4.3.3.1 Effects of Aspirin and Other Pain Killers: Prostaglandins induce inflammation, pain, and fever. Aspirin blocks an enzyme called cyclooxygenase, which is involved with the ring closure and addition of oxygen to arachidonic acid converting to prostaglandins. The acetyl group on aspirin is transferred to the alcohol group of serine as an ester. This has the effect of blocking the channel in the enzyme and arachidonic can not enter the active site of the enzyme.

By inhibiting or blocking this enzyme, the synthesis of prostaglandins is blocked, which in turn relieves some of the effects of pain and fever.

Aspirin also inhibits the prostaglandin synthesis involved with unwanted blood clotting in coronary heart disease. At the same time during an injury taking aspirin may cause more extensive bleeding.

Thromboxanes are labile transformation products of certain prostaglandins. Thromboxanes A_2 (TXA_2) causes aggregation of platelets and has the opposite effect as compared to Prostacyclins (PGI_2).

Leukotrienes was first found in the leukocytes. However, they are also formed in macrophages and other tissues such as lung, spleen, brain, heart in response to immunological and non-immunological response. Leukotriene A is a precursor to other leukotriene derivatives by epoxide opening reactions. Leukotrienes are powerful biological signals. Leukotriene D_4 derived from Leukotrienes A_4 induces contraction of the muscle lining the airways to the lung. Overproduction of leukotrienes causes asthamatic attacks, and leukotriene synthesis is one target of anti-asthamatic drugs.

5.4.3.4 Terpenes

The terpenes have been long associated with the term Essential Oils comprising resins, steroids and rubber. They are hydrocarbons that usually contain one or more C=C double bonds, while the terpenoids are oxygen-containing analogues of the terpenes. They are thoroughly distributed only in the plant kingdom, especially in those plants that have abundant chlorophyll, but some of the larger and more complex terpenes (*e.g.*, squalene & lanosterol) occur in animals.

Compounds which fall in the general class of terpenes, compounds made up of 5-carbon units, often called isoprene units (2-methyl-1, 3-butadiene), are joined together in a regular pattern, usually head-to-tail in terpenes up to 25 carbons. Terpenes containing 30 carbons or more are usually formed by the fusion of two smaller terpene precursors such that the heat-to-tail "rule" appears to be violated. Overall, terpenes hold potential interest in practical applications especially in the fragrance and flavour industries, as well as in the pharmaceutical and chemical industries.

$$CH_2{=}C(CH_3){-}CH{=}CH_2$$

Isoprene

5.4.3.4.1 Classification: Depending on the number of 5-carbon units, the terpenes are subdivided into several groups.

Classification		*Isoprene units*	
Hemiterpenes	C5H8	1	Constituent of mixed terpene. Consists of a terpenoid and a non-terpenoid components. Found in the side chains of quinones.
Monoterpenes	C10H16	2	They have either acyclic (myrcene, geraniol, citral, citronellol) or cyclic structure (either monocyclic — carvone limonene, menthol, menthone, thymol or bicyclic — α and β-pinene, camphor).
Sesquiterpenes	C15H24	3	They have either acyclic (farnesol) or cyclic structure (abscisic acid).
Diterpenes	C20H32	4	They have either acyclic (phytol) or cyclic structure (gibberellin).
Sesterterpenes	C25H40	5	
Triterpenes	C30H48	6	They have either acyclic (squalene) or cyclic structure (steroids).
Tetraterpenes	C40H64	8	Consists of carotene and xanthophyll.
Polyterpenes	(C5H8)n		Consists of isoprene units — natural rubber, cis — configuration, gutta-percha, trans configuration.

Isoprene itself, a C_5H_8 gaseous hydrocarbon, is emitted by the leaves of various plants as a natural byproduct of plant metabolism. Next to methane it is the most common volatile organic compound found in the atmosphere.

Monoterpenes

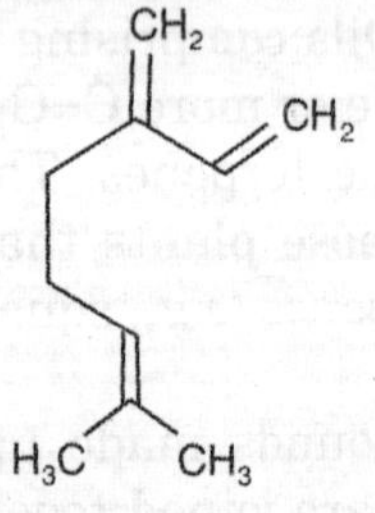

CH_2 CH_2 H_3C CH_3

myrcene

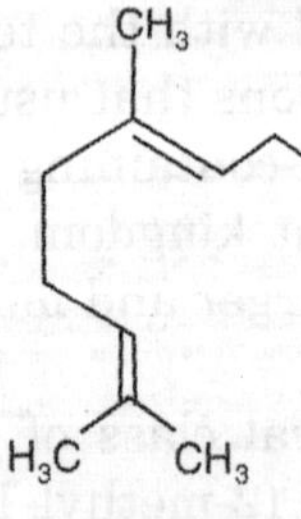

CH_3 OH H_3C CH_3

geraniol

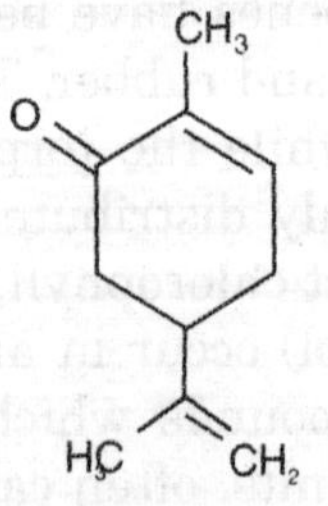

CH_3 O H_3C CH_2

carvone

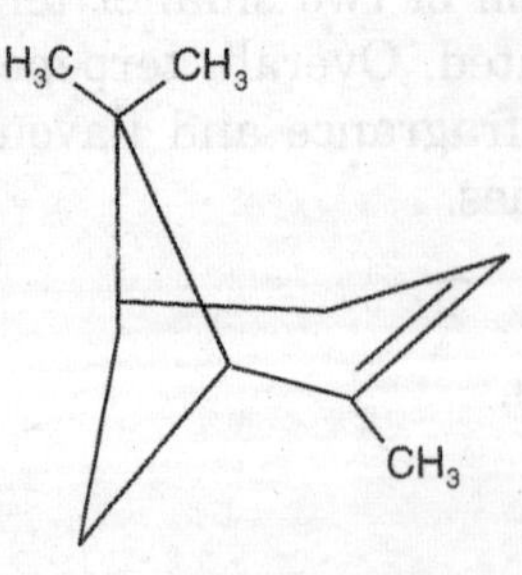

H_3C CH_3 CH_3

α-pinene

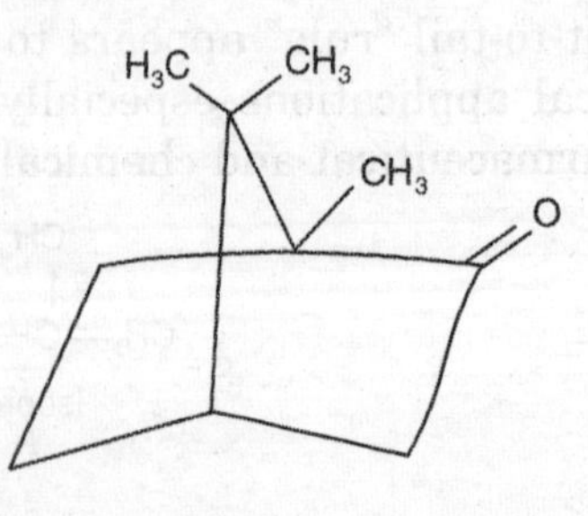

H_3C CH_3 CH_3 O

camphor

Polymeric isoprenoid hydrocarbons have also been identified. Rubber is the best known and most widely used compound of this kind. It occurs as a colloidal suspension called latex in a number of plants, ranging from the dandelion to the rubber tree (*Hevea brasiliensis*). Rubber is a polyene, and exhibits all the expected reactions of the C=C function. Bromine, hydrogen chloride and hydrogen, all add with a stoichiometry of one molar equivalent per isoprene unit. Pyrolysis of rubber produces the diene isoprene along with other products.

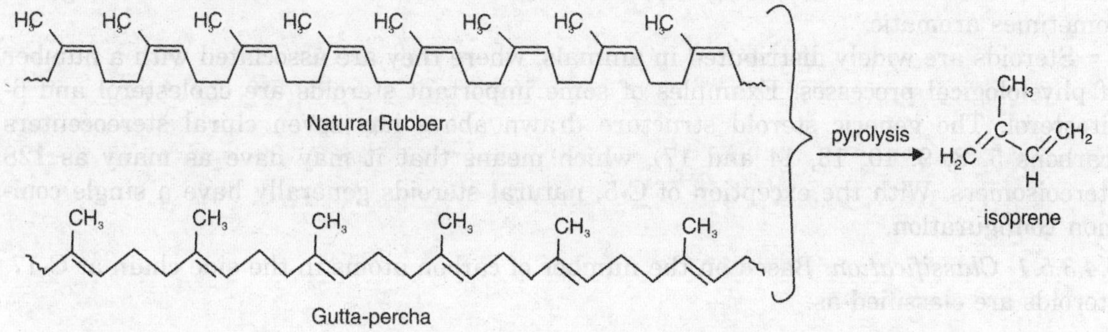

Natural Rubber

pyrolysis

isoprene

Gutta-percha

The double bonds in rubber — all have a **Z**-configuration (cis-configuration), which causes this macromolecule to adopt a kinked or coiled conformation. This is reflected in the physical properties of rubber. Despite its high molecular weight (about one million), crude latex rubber is a soft, sticky, elastic substance. Chemical modification of this material is normal for commercial applications. Gutta-percha, is a naturally occuring **E** — isomer (trans-isomer) of rubber. Here the hydrocarbon chains adopt a uniform zig-zag or rod-like conformation, which produces a more rigid and tough substances. Uses of gutta-percha include electrical insulation and the covering of golf balls.

5.4.3.5 Steroids

The important class of lipids called steroids are actually metabolic derivatives of terpenes, but they are customarily treated as a separate group. They are non-saponifiable lipids having specialised functions. These are the most studied classes of biological compounds because of their physiological activity.

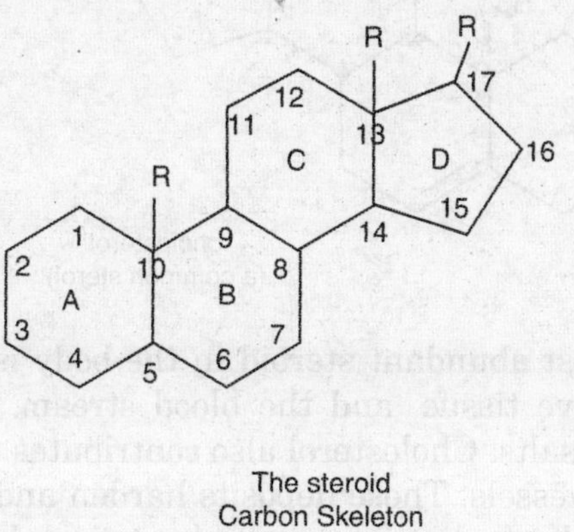

The steroid
Carbon Skeleton

Steroids may be recognised by their tetracyclic skeleton, consisting of three fused six-membered and one five-membered ring. The four rings are designated A, B, C & D, and the peculiar numbering of the ring carbon atoms is the result of an earlier misassignment of the structure. The substituents designated by R are often alkyl groups, but may alsohave functionality. The R group at the A:B ring fusion is most commonly methyl or hydrogen, that at the C:D fusion is usually methyl. The substituent at C-17 varies considerably, and is usually larger than methyl if it is not a functional group. The most common locations of functional groups are C-3, C-4, C-7, C-11, C-12 and C-17. Ring A is sometimes aromatic.

Steroids are widely distributed in animals, where they are associated with a number of physiological processes. Examples of some important steroids are cholesterol and β-sitosterol. The generic steroid structure drawn above has seven chiral stereocenters (carbons 5, 8, 9, 10, 13, 14 and 17), which means that it may have as many as 128 stereoisomers. With the exception of C-5, natural steroids generally have a single common configuration.

5.4.3.5.1 Classification: Based on the number of carbon atoms in the side chain at C-17, steroids are classified as:

- Sterols, containing 8–10 Carbon atoms in the side chain at C-17
- Bile acids, containing 5 Carbon atoms in the side chain at C-17
- Adrenocortical hormones and progestins, containing 2 Carbon atoms in the side chain at C-17
- Estrogen and androgens, containing no carbon atoms in the side chain at C-17

5.4.3.5.1.1 Sterols: The first steroid isolated from nature were a series of C27–C29 alcohols found in the lipid fractions of many tissues. These compounds were solids and were therefore named Sterols (in greek, stereos means solid). Steroids with 8–10 carbon atoms in the side chain at C-17, hydroxyl group at C-3 are called Sterols. The most abundant sterol in animal tissue is cholesterol and in plant — sitosterol.

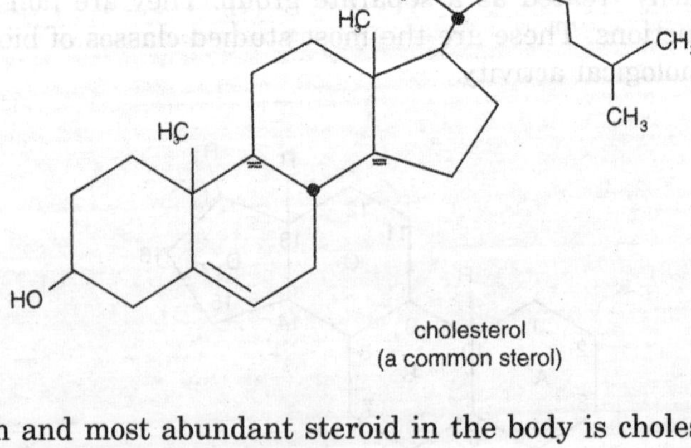

H_3C H_3C CH_3 CH_3 H_3C HO

cholesterol
(a common sterol)

The best known and most abundant steroid in the body is cholesterol. Cholesterol is formed in brain tissue, nerve tissue, and the blood stream. It is the major compound found in gallstones and bile salts. Cholesterol also contributes to the formation of deposits on the inner walls of blood vessels. These deposits harden and obstruct the flow of blood. This condition, known as atherosclerosis, results in various heart diseases, strokes, and high blood pressure. Much research is currently underway to determine if a correlation exists between cholesterol levels in the blood and diet. Not only does cholesterol come from the diet, but cholesterol is synthesized in the body from carbohydrates and proteins as well as fat. Therefore, the elimination of cholesterol rich foods from the diet does not necessarily lower blood cholesterol levels. Some studies have found that if certain unsaturated fats and oils are substituted for saturated fats, the blood cholesterol level decreases.

Cholesterol is derived from acetate. Three molecules of acetate, derived metabolically combine to form isoprene — a 5-Carbon molecule, which then is used to compose squalene — a size unit isoprene molecule. Squalene can cyclisize and transform to form the 27 carbon structure of cholesterol. Cholesterol is very important, in the production of steroid hormones. They are the precursor for bile acids (bile acids aid in fat digestion), steroid hormones, and provitamin D (When irradiated by sunlight it changes to vitamin D3).

Cholesterol, is incorporated into the cell membrane by lipoproteins. There it plays a role in the regulation of membrane fluidity. It has been stated that cholesterol is probably responsible for permitting steroid hormones to enter the cell.

The most common sterol in plant is sitosterol, which is widely distributed throughout the plant kingdom.

5.4.3.5.1.2 Bile Acids: Steroids containing 5-Carbon atoms in the side chain at C-17 and terminating in a carboxyl group are called bile acids. These have been isolated from human bile.

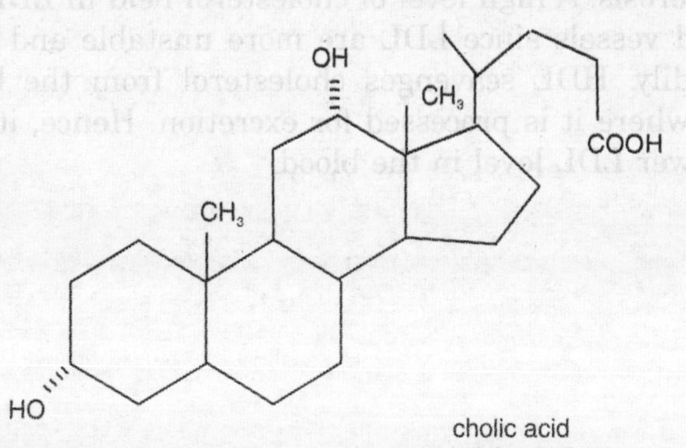

OH
CH_3
COOH
CH_3
HO

cholic acid
(a bile acid)

In the liver of man and other animals, the side chain of cholesterol is degraded to C-24 steroids containing a terminal carboxyl group at C-24. These are collected in the bile and are referred as bile acids. The primary bile acid formed in the liver is cholic acid. Generally, bile acids do not exist in the free form, but are conjugated to either glycine or taurine to form glycocholic acid and taurocholic acid, and are discharged into the duodenum. The sodium salts of the conjugated bile acids, act as emulsifying agents, and aid in the intestinal absorption of fats.

5.4.4 Lipids Combined with other Classes of Compounds

Some of the lipids are associated with specific proteins to form lipoprotein systems. They have the solubility characteristics of proteins (water soluble). There are two major types — Transport lipoproteins and Membrane lipoproteins. In these systems, the lipids and proteins are not covalently joined but are held together largely by hydrophobic interactions between the nonpolar portions of the lipid and the protein components.

The plasma lipoproteins are complexes in which the lipids and proteins occur in a relatively fixed ratio and are classified on the basis of their lipid content into four major classes, which is reflected in their size and density — chylomicrons (density 0.94 g/ml), very low density lipoprotein (VLDL) or pre-lipoprotein (density 0.94–1.006 g/ml), low density lipoprotein (LDL) or β-lipoprotein (density 1.006–1.063 g/ml), and high density lipoprotein (HDL) or α-lipoprotein (density 1.063–1.21 g/ml). The greater their lipid content, the lower their density and the greater their tendency to move upward when blood plasma is centrifuged at very high speed. They carry water insoluble lipids between various organs via the blood.

Chylomicrons (75–1000 nm) are droplets of nearly pure triacylglycerides surrounded by a very thin layer of protein (1–2%) and are much larger than lipoproteins.

VLDL (30–50 nm) contain predominantly triacylglycerides, while LDL (20–22 nm) and HDL (7.5–10 nm) contain predominantly cholesterol and phospholipids, respectively. VLDL, LDL and HDL contain about 10%, 25% and 45–55% protein, respectively. The presence of a high level of plasma LDL with a low level of plasma HDL results in atherosclerosis — the formation of thick deposits of cholesterol and its esters on the inner surface of blood vessels. HDL picks up cholesterol from the plasma and do not release cholesterol into the plasma since HDL are very stable. A high level of HDL lessen the incidence of atherosclerosis. A high level of cholesterol held in LDL leads to deposition in the walls of the blood vessels since LDL are more unstable and release the cholesterol into the plasma readily. HDL scavenges cholesterol from the blood and tissues and delivers to the liver, where it is processed for excretion. Hence, it is desirable to have a higher HDL and a lower LDL level in the blood.

6

Vitamins

Vitamins are generally defined as organic molecules present in the diet in small amount to execute unique biological functions for growth and development and health of the organisms. Higher organisms can not synthesise in the body, therefore vitamins are dependent on dietary supply for small amount of vitamins. Most of the vitamins are present in the diet as intact molecules or few are present as precursors known as provitamins. Once these are ingested inside the body it can be converted into vitamins.

6.1 CLASSIFICATION OF VITAMINS

Vitamins are classified into fat-soluble and water-soluble vitamins. Fat-soluble vitamins include vitamin A, D, E and K. Whereas water-soluble vitamins are C and B group. The water-soluble vitamin group is once again sub-classified into energy-releasing (B_1, B_2, B_6, biotin and pantothenic acid) and hematopoietic (folic acid or vitamin B_{12}) vitamins.

6.1.1 Fat-Soluble Vitamins

Fat-soluble vitamins such as vitamin A, D, E and K are lipid soluble and require bile salts for absorption. They are stored in liver and adipose tissue. They are not readily excreted in urine, instead excreted through faeces.

6.1.1.1 Vitamin A

6.1.1.1.1 Chemistry: Chemically, vitamin A is a group of compounds which shows vitamin A activity. The term retinoids is often used to include the natural and synthetic forms of vitamin A. Vitamin A compounds are retinol, retinal and retinoic acid. Retinol is basically an alcohol containing β-ionic ring and its side chain has two isoprenoid units. Presence of double bonds is responsible for oxidation. Retinal is an aldehyde form produced by retinol. Similarly, retinoic acid is produced by the oxidation of retinal. β-carotene, a preculator for vitamin A is found in plant foods. It undergoes cleavage process to produce two moles of retinal. However, this conversion is inefficient.

6.1.1.1.2 Absorption and transport: In the intestine, dietary retinyl esters are hydrolysed by hydrolases releasing retinol and free fatty acids. Carotenes are hydrolysed by β-carotene dioxygenase to release 2 moles of retinal which is then reduced to retinol. The intestinal mucosal cells restricted retinol to long chain fatty acids and then incorporated to chylomicrons transferred to lymph. The retinol esters of chylomicrons are uptaken by liver and then stored.

6.1.1.1.3 Dietary Source: It is present as retinol esters in animal foods, codliver fish oil, shark liver oils and halibut liver oils are excellent source. In addition, egg, milk and butter are good source. Among plants, it is present as carotenes in fruits and vegetables, plant oil like red palm oil is an excellent source. Many leafy vegetables such as coriander, curry leaves, drumstic leaves and spinach are excellent source.

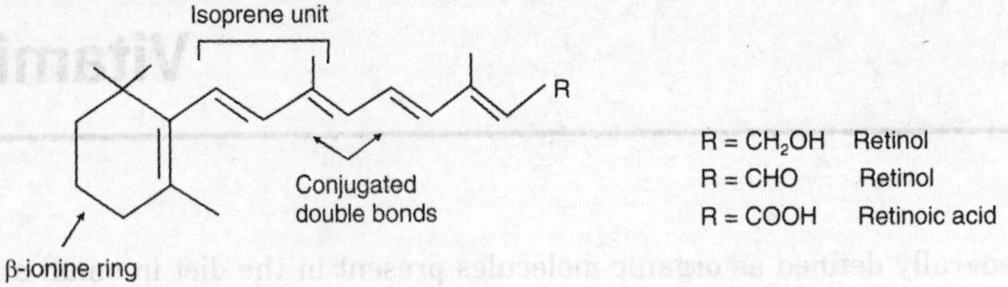

Isoprene unit
R
Conjugated double bonds
β-ionine ring
R = CH_2OH Retinol
R = CHO Retinol
R = COOH Retinoic acid

Structure of Vitamin A

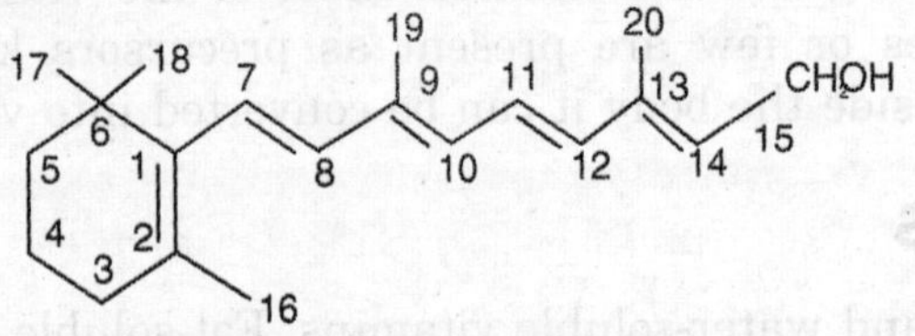

17
18
19
20
7
9
11
13
CH_2OH
6
15
5
1
8
10
12
14
4
2
3
16

All-trans Retinol

$C_{20}H_{30}O$

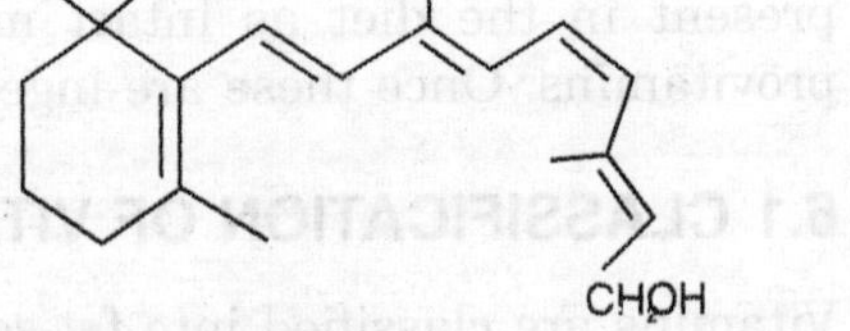

CH_2OH

11-cis-Retinol

Geometric isomers of Retinol (Vitamin A)

Retinal = Vitamin A alcohol.
Retinal = Vitamin A aldehyde.
Retinoic acid = Vitamin A acid.

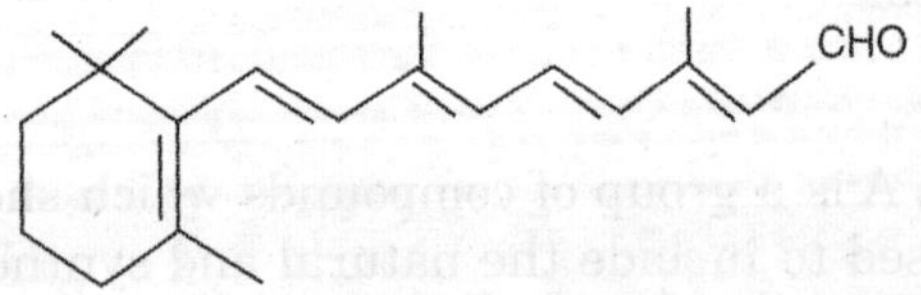

CHO

2 Retinal

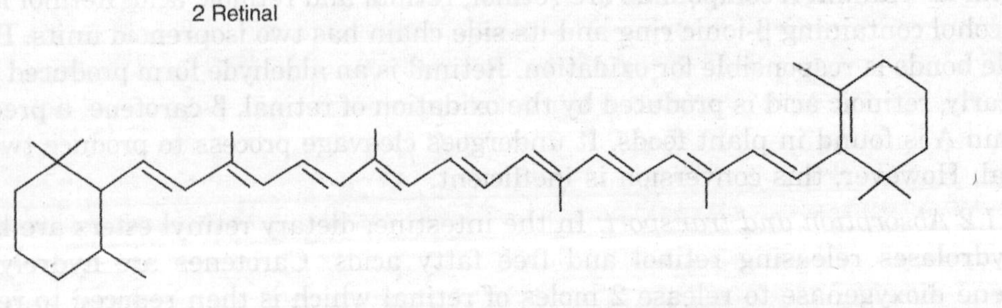

Structure of βCarotene (Provitamin of Vitamin A) Plant origin.

6.1.1.1.4 Functions of vitamin A

1. Retinol is essential for healthy colour vision.
2. Retinol is indispensable for reproduction and growth as it facilitates spermatogenesis and Oogensis.
3. Vitamin A is required for healthy skin, respiratory tract and urinary tract.
4. Retinoic acid acts as steroid hormone thereby it promotes growth and differentiation.
5. Vitamin A is also essential for tooth formation and bone growth.

6.1.1.2 Vitamin D

6.1.1.2.1 Chemistry: There are two groups of compounds that exhibit vitamin D activity. These are vitamin D_2 (calciferol) and vitamin D_3 (cholecalciferol). These two vitamin D forms are formed from provitamins like sterols. The ergosterol found in ergol and yeast acts as provitamin for vitamin D_2 which is actually a derivative of cholesterol. Similarly, the 7-dehydrocholesterol, which is found in animals acts as provitamin of vitamin D_3.

Vitamin D

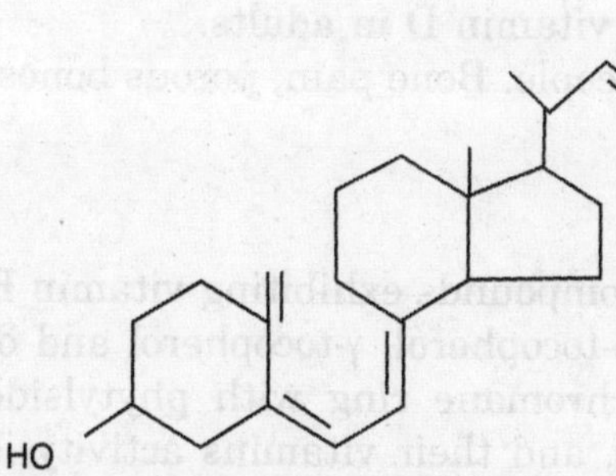

HO

Ergo calciferol (Vitamin D_2)
Structure of Vitamin D_2

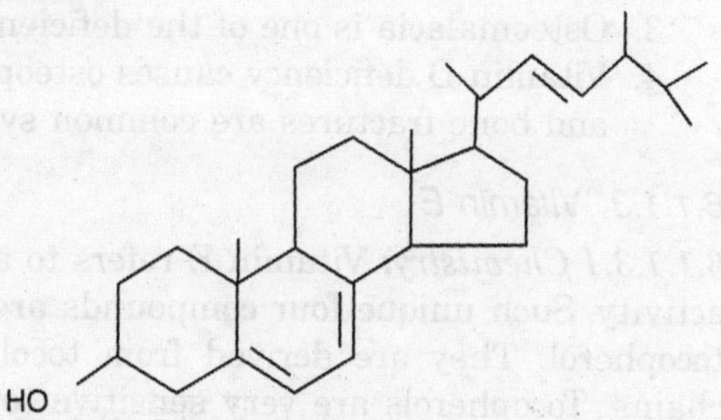

HO

Ergosterol (Provitamin of Vitamin D_2)
Structure of Provitamin of Vitamin D_2

CH₂

HO

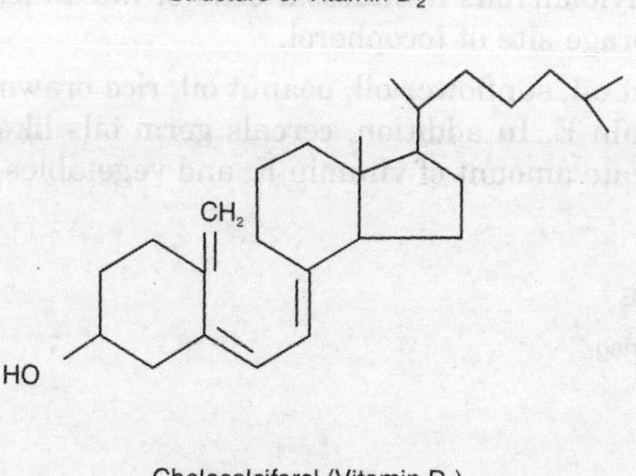

Cholecalciferol (Vitamin D_3)
Structure of Vitamin D_3

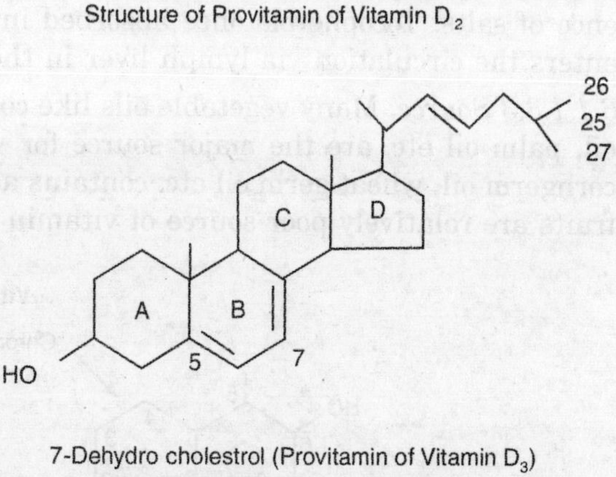

26 25 27 C D A B 5 7 HO

7-Dehydro cholestrol (Provitamin of Vitamin D_3)
Structure of Provitamin of Vitamin D_3

Both vitamin D_2 and D_3 are popularly called as sunshine vitamins, since their formation is dependent on sunlight. The above said provitamins are converted into active forms on exposure to ultraviolet in the sunlight. In humans, 7-dehydrocholesterol — a provitamin D_3, present below the skin is converted to vitamin D_3 on exposure to sunlight.

6.1.1.2.2 Absorption and transport: The vitamin D_2 and D_3 are absorbed in the small intestine in presence of bile salts. Vit D is influxed into chylomicrons which then enters circulation via the lymph. Vitamin D can dissociate from chylomicrons during circulation and binds to specific vitamin D binding (DBP) proteins. Finally various types of tissues take up vitamin D from DBP. Vitamin D is stored in the liver as adipose tissue.

6.1.1.2.3 Source: It is abundantly present in most of the foods of animal origin. Certain marine fish liver oils such as halibut liver oil, cod liver oil and shark liver oil are good source. In addition, egg yolk and butter contains meagre amount of vitamin D. Mushroom contains little amount of vitamin D.

The daily requirement of vitamin D is 200 international units (IU)/day or 5 μg of vitamin D.

6.1.1.2.4 Deficiency symptoms

1. Vitamin D deficiency causes rickets among children. Rickets among children are bone deformities because of incomplete mineralization.
2. Deficiency leads to delay in teeth formation.
3. Osteomalacia is one of the deficiency symptoms of vitamin D in adults.
4. Vitamin D deficiency causes osteoporosis in aged people. Bone pain, porous bones, and bone fractures are common symptoms.

6.1.1.3 Vitamin E

6.1.1.3.1 Chemistry: Vitamin E refers to a group of four compounds exhibiting vitamin E activity. Such unique four compounds are α-tocopherol, β-tocopherol, γ-tocopherol and δ-tocopherol. They are derived from tocol or 6-hydroxy chromane ring with phytylside chains. Tocopherols are very sensitive towards alkalinity, and their vitamins activity is destroyed by oxidation. α-tocopherol is widely distributed in nature.

6.1.1.3.2 Absorption and transport: Small intestine absorbs dietary tocopherols in presence of salts. Tocopherols once absorbed into chylomicrons in mucosal cells of intestine, enters the circulation via lymph liver in the storage site of tocopherol.

6.1.1.3.3 Source: Many vegetable oils like coconut oil, sunflower oil, peanut oil, rice brawn oil, palm oil etc. are the major source for vitamin E. In addition, cereals germ oils like corngerm oil, wheat germ oil etc. contains adequate amount of vitamin E, and vegetables, fruits are relatively poor source of vitamin E.

Vitamin E

Chromane ring

HO

Methyl groups

Phytyl side chain

Structure of α-tocopherol (3 methyl groups at positions 5, 7 & 8 of chromane ring).

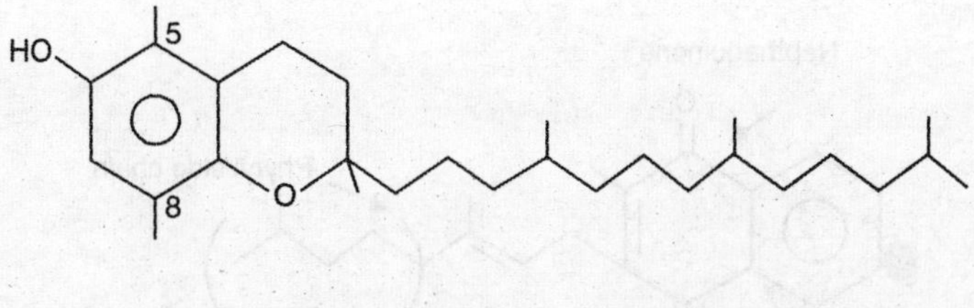

HO 5 8 O

Structure of βtocopherol (2 methyl groups at positions 5 & 8 of chromane ring).

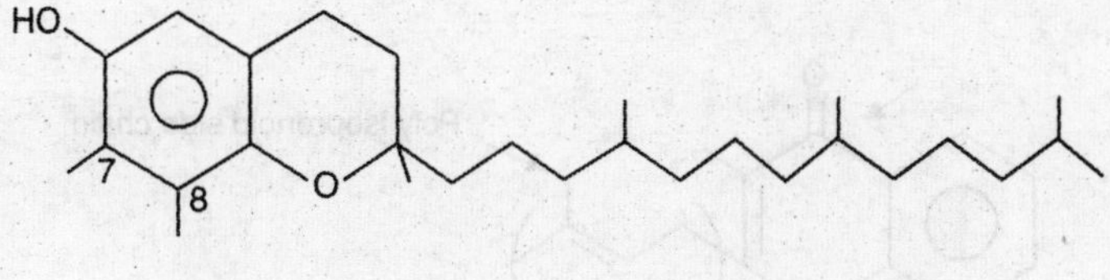

HO 7 8 O

Structure of γtocopherol (2 methyl groups at positions 7 & 8 of chromane ring).

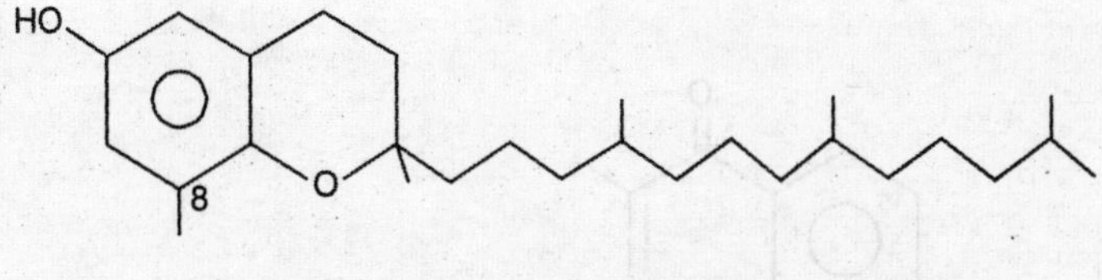

HO 8 O

Structure of δtocopherol (1 methyl group at position 8 of chromane ring).

Structures of tocopherols (Vitamin E)

Polyunsaturated fatty acid (PUFA) supplements the daily requirement of vitamin E. The adults needs 10 mg of vitamin E per day.

6.1.1.3.4 Functions: α-tocopherol acts as antioxidant. It precludes peroxidation of membrane lipids. It protects the unsaturated fatty acids (PUFA) from peroxidation reactions. It is essential for reproduction function. Vitamin E prevents oxidation of vitamin A and carotenes. It is involved in the biosynthesis of nucleic acids.

High doses of vitamin E are employed in treating cardiovascular diseases, sterility and muscular dystropy and even diabetes. Ageing can be prevented by administering adequate amount of vitamin E.

6.1.1.4 *Vitamin K*

Chemistry: There are groups of vitamin K compounds which shows vitamin K activities. They are vitamin K group (phylloquinone), found in green leafy vegetables, and vitamin K_2 group (Menaquinone) found in animals, especially synthesised by intestinal flora. They are derivatives of napthaquinone. Phylloquinone contains phytyl side chain, where as menaquinone contains polyisoprenoid side chain.

Vitamin K

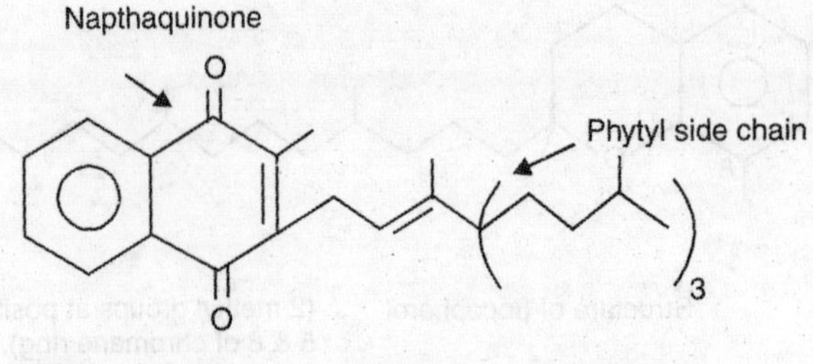

Napthaquinone

Phytyl side chain

Structure of Vitamin K_1 (Phylloquinone, plant origin).

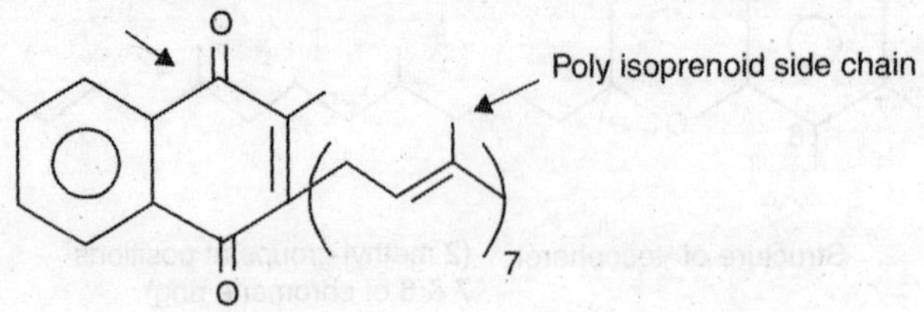

Poly isoprenoid side chain

Structure of Vitamin K_2 (Menaquinone, animal origin).

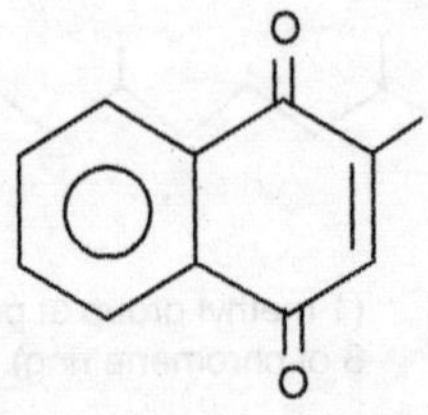

Structure of Menadione (Vitamin K_3 Synthetic analog of Vitamin K).

Absorption and Transport: In presence of bile salts, dietary vitamin K get absorbed in small intestine. The absorbed vitamin K is incorporated into chylomicrons, which later enters the circulation through the lymph and is stored in liver.

6.1.1.4.1 Functions:

1. It is required for the synthesis of blood clotting factors prothrombin (factor 11), proconvertion or cothromboplastin (factor VII) and chrismas factor (factor X).
2. It is essential for the carboxylation of the γ-carbon atom of glutamic residue of clotting factors.

6.1.1.4.2 Source: Plant sources like peas, soyabean, cauliflower and cabbage are nice sources for vitamin K. Cereals such as rice, wheat, tomatoes, peaches, and banana contains moderate amount.

Dairy products like cheese, butter and egg contains adequate amount of vitamin K. Generally, 70–140 mg/day is needed for daily requirement.

6.1.2 Water-Soluble Vitamins

6.1.2.1 Vitamin C (Ascorbic acid)

6.1.2.1.1 Chemistry: Vitamin C is sugar acid called as hexuronic acid. It is easily oxidised by O_2 to dehydroascorbic acid. Its oxidation enhances at high temperature during cooking. This vitamin plays a very important role in human health.

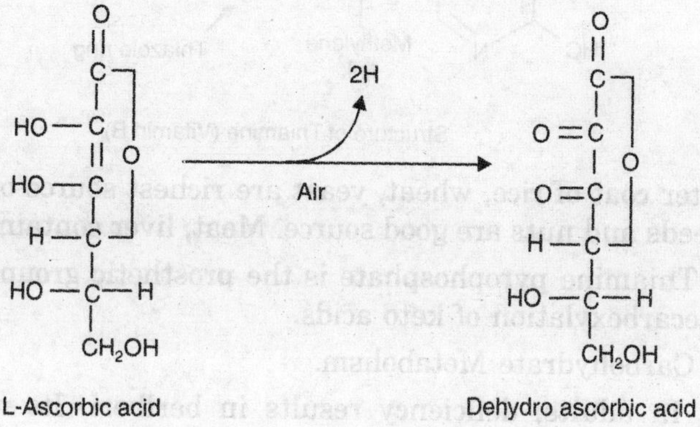

L-Ascorbic acid —(Air, −2H)→ Dehydro ascorbic acid

Ascorbic acid (Vitamin C)

6.1.2.1.2 Functions

- It is a good antioxidant.
- It facilitates bile acid synthesis in liver.
- Participate in hydroxylation of steroid biosynthesis reaction.
- Used in the treatment of cold.

6.1.2.1.3 Source: Mainly available in citrus formite, gooseberry, guava, green vegetables like tomatoe, potatoe and cabbage. Milk is a poor source of ascorbic acid.

6.1.2.1.4 Deficiency: Vitamin C deficiency in adults results in scurvy—symptoms of scurvy are hemorrhages in tissues, weakness and anaemia.

Its daily requirement in adults is 60–80 M/day.

6.1.3 Vitamin B Complex

The important members of vitamin B complex are:

1. Thiamine (vitamin B_1)
2. Riboflavin (vitamin B_2)
3. Niacin
4. Pyridoxine (Vitamin B_6)
5. Biotin
6. Folic acid
7. Cyanocobalamine (B_{12})
8. Pantothenic acid

6.1.3.1 Thiamine

6.1.3.1.1 Chemistry: It is sulphur containing partially heat labile vitamin. Prominent feature is the presence of pyrimidine and thiazole ring joined by methylene bridge.

Structure of Thiamine (Vitamin B)$_1$

6.1.3.1.2 Source: Outer coat of rice, wheat, yeast are richest source of vitamin B_1. Whole cereals, pulses, oil seeds and nuts are good source. Meat, liver contains moderate amount.

6.1.3.1.3 Functions: Thiamine pyrophosphate is the prosthetic group of enzymes participated in oxidative decarboxylation of keto acids.

– Essential for Carbohydrate Metabolism.

6.1.3.1.4 Deficiency: In adults, deficiency results in beriberi. Its main symptoms are insomnia.

– In another type of beriberi known as wet beriberi: cardiovascular system is affected. In dry beriberi central nervous system is affected.

Among infants, thiamine deficiency causes infantile beriberi.

6.1.3.1.5 Daily requirement: The daily requirement of thiamine depends on the carbohydrate content of food as it involves in carbohydrate metabolism.

The daily intake of 0.4 mg/1000c mg been recommended of dietary source is carbohydrate. Hard working groups require 2 mg per day.

6.1.4.1 Riboflavin

6.1.4.1.1 Chemistry: Riboflavin contains heterocyclic isoalloxazine ring and ribitol — a sugar alcohol. It is stable to heat and acidic pH.

Structure of Riboflavin Vitamin B_{12}

6.1.4.1.2 Functions: Riboflavin is in the form of FMN and FAD and acts as prosthetic group of many enzymes. The flavoenzymes catalyse oxidation and reduction reactions.

6.1.4.1.3 Sources: Green leafy vegetables, pulses, eggs and milk are the good sources of riboflavin. Fruits and some root vegetables contain moderate amount.

6.1.4.1.4 Deficiency: In humans it causes oral, facial and occular lesions. In experimental animals, riboflavin deficiency resulted in growth retardation, cataract formation and corneal problems.

6.1.4.2 Niacin

Niacin is a derivative of two pyridine derivatives. They are nicotinic acid and nicotinamide and are stable to heat and alkali.

6.1.4.2.1 Sources: Legumes, yeast, fish, liver, whole grains and meat are good sources.

6.1.4.2.2 Functions: Nicotinamide is a component of two co-enzymes NAD and NADP, which catalyse oxidation and reduction. The pyrimidine ring participate in reduction of substrate.

Pyridine ring

COOH

N

Nicotinic acid

$CONH_2$

N

Nicotinamide

Structures of nicotinic acid and nicotinamide
(The word Niacin refers to these two pyridine derivatives)

6.1.4.2.3 Deficiency

- It causes pellegra.
- In pellegra, skin, gastrointestinal tract and nervous system are affected, therefore, dermatitis, diarrhea are main symptoms of pellegra.
- Glossitis and stomatitis are also other symptoms of niacin deficiency.
- *Pyridoxine:*

  Certain compounds derived from pyridine like pyridorine, pyridoxal and pyridoxamin show vitamin B_6 activity. These compounds are stable to heat and sensitive to light and alkali.

CH_2OH

HO

CH_2OH

H_3C

N

Pyridoxine

CHO

HO

CH_2OH

H_3C

N

Pyridoxal

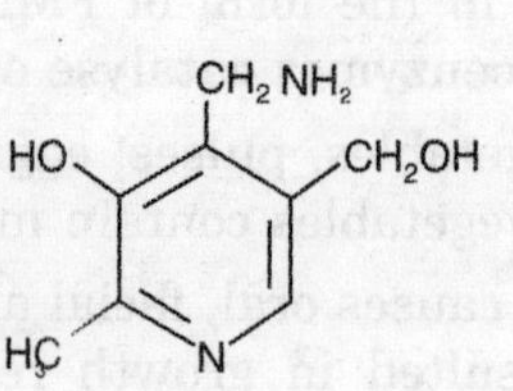

Structure of Vitamin B_6

6.1.4.2.4 Source: **Pulses, liver, whole grains and yeast contain adequate amount of pyridoxine. Leafy vegetables, milk, meat and eggs also contain moderate level of pyridoxine.**

6.1.4.2.5 Functions: **Pyridoxal phosphate act as the prosthetic group or coenzyme involved in transamination, decarboxylation and non-oxidative deamination reactions.**

6.1.4.2.6 Deficiency

- In children, deficiency causes epileptic form convulsions
- Growth retardation
- Microcytic hypochromic anaemia
- Skin lesions

6.1.4.3 Biotin

It is a sulphur containing vitamin and consist of imidazole ring. It is stable to heat but sensitive to alkaline medium.

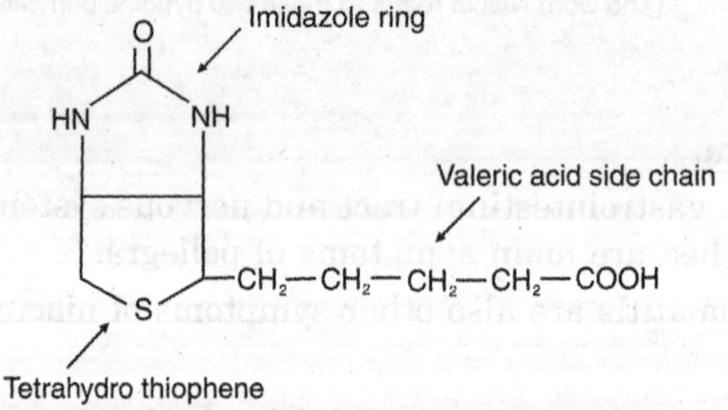

Structure of Biotin

6.1.4.3.1 Source: Green leafy vegetables like cabbage, spinach, mint leaves, pulses, eggs and liver are good sources. Milk, whole cereals contains adequate amount.

6.1.4.3.2 Function: Biotin is the prosthetic group of many enzymes like pyruvate carboxylase, acetyl-CoA carboxylase, etc. It involves in carrier of CO_2 in carboxylation reaction.

6.1.4.3.3 Deficiency

- Deficiency causes dermatitis, depression, muscular pain and anaemia.
- In experimental animals, deficiency causes several neurological problems.

6.1.4.4 Folic Acid

It consists of pteridine nucleus, p-aminobenzoic acid and glutamate. Folic acid is sensitive to light and acid, but stable to heat.

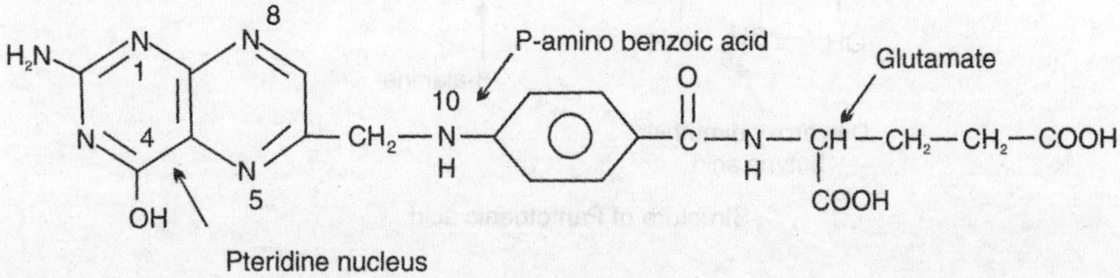

Pteridine nucleus — P-amino benzoic acid — Glutamate

Structure of Folic acid

6.1.4.4.1 Sources: Green leafy vegetables, pulses like green gram, black gram, bengal gram and eggs are adequate sources of folid acid. Whole cereals, coconuts contain adequate source of folic acid. Folic acid acts as the carrier of one carbon units.

6.1.4.4.2 Deficiency: In humans, type of anaemia known as megaloblastic anaemia is the major symptom of folic acid deficiency particularly in pregnant women and children. Generally, rapidly dividing cells like bone marrow or intestinal cells are most affected. Other symptoms of folid acid deficiency are leucopenia, diarrhea and weakness.

In daily requirement, folic acid are 0.1 mg/day. Whereas pregnant women requires 0.15 mg/day.

6.1.5.1 Cyanocobalamin (Vitamin B_{12})

Cyanocobalamin is made up of tetra pyrrole ring system known as corrin ring with a central cobalt atom. It also contains unusual nucleotide in which the nitrogenous base 5, 6-dimethyl benzinidazole. The R^- group is attached to the central cobalt atom. Although many forms of vitamin B_{12} are named based on 'R' group attachment to the central cobalt atom, most of the therapeutic preparations contain cyanocobalamin.

6.1.5.1.1 Sources: Animal sources like kidney, liver, brain, fish and eggs are good sources. Plants are devoid of vitamin B_{12}. Daily requirement of B_{12} is 3–4 mg/day for adults.

6.1.5.1.2 Functions: Vitamin B_{12} acts as the prosthetic group or coenzyme.

6.1.5.1.3 Deficiency:

- Affects bone marrow and neurological system due to alteration in fatty acid and DNA synthesis.
- Methionine synthesis is blocked.
- Deficiency symptoms are sore tongue, megaloblastic anaemia, degeneration of spinal cord and malabsorption.

6.1.6.1 Pantothenic Acid

Chemically, it is amide of β-alanine and dihydroxy dimethyl butyric acid. It is unstable to acid or alkali. It is however, stable to heat.

6.1.6.1.1 Sources: Cereals, legumes, oxygen meat, liver, milk and eggs are good source.

Daily requirement is 4–6 mg/day for adults.

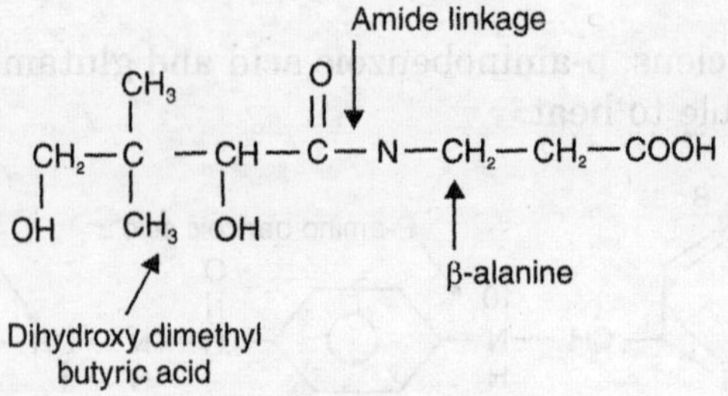

Structure of Pantothenic acid

6.1.6.1.2 Functions:

- It is a component of coenzyme A.
- It is required for the synthesis of phosphopantotheine of fatty acid synthase complex.

6.1.6.1.3 Deficiency:

- Rare in humans, experimental animals produces paristhesia of extremities *i.e.*, burning feet, abdominal cramps, fatigue.
- Dermatitis, graying of hair, growth retardant and neurological lesions.

7

Hormones

7.1 INTRODUCTION

Steroid hormones are very important substances required for the proper functioning of the body. They mediate a wide variety of vital physiological functions ranging from anti-inflammatory agents to regulating events during pregnancy. They are synthesised and secreted into the bloodstream by endocrine glands such as the adrenal cortex and the gonads (ovary and testis). Steroid hormones are all characterised by the steroid nucleus which is composed of three six member rings and one five member ring, labeled A, B, C and D respectively.

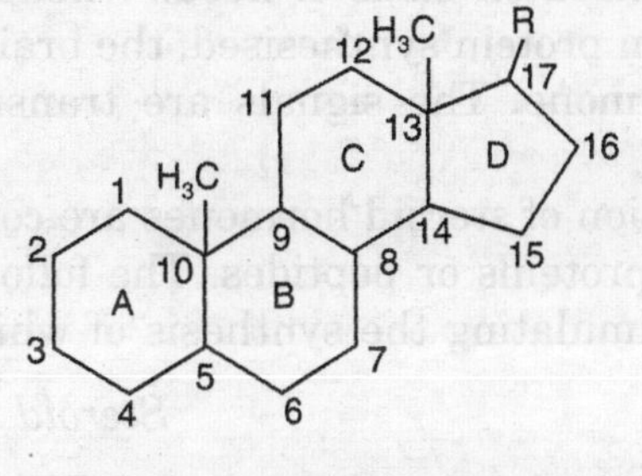

A steroid
(R - various side chains)

The steroid nucleus is known as cyclopentanophenanthrene. This structure, which has six asymmetric carbons, provides many possible stereo isomers as one would expect since steroid hormones have an array of functions. Furthermore at C-17, there is a substituent which varies from hormone to hormone, depending on its function.

The steroid hormones are all derived from cholesterol. Moreover, with the exception of vitamin D, they all contain the same cyclopentanophenanthrene ring and atomic numbering system as cholesterol. The conversion of C_{27} cholesterol to the 18-, 19-, and 21-carbon steroid hormones (designated by the nomenclature C with a subscript number indicating the number of carbon atoms, *e.g.*, C_{19} for androstanes) involves the rate-limiting, irreversible cleavage of a 6-carbon residue side chain from cholesterol, producing pregnenolene (C_{21}) plus isocaproaldehyde.

All hormones have an oxygen at C-3 and a varied substituent at C-17. This substituent varies according to the different kind of steroid hormones and can be either α-, β-, depending on whether they are situated below the plane of the molecule or above the plane. There are 6-centers of asymmetry, as a result, there are 64 possible compounds (stereoisomers) with this structure. They are located at C-3, C-8, C-9, C-10, C-13 and C-17.

Steroids with 21 carbon atoms are known systematically as pregnanes, whereas those containing 19 and 18 carbon atoms are known as androstanes and estranes, respectively. The important precursor of mammalian steroid hormones are the pregnenolone.

Pregnenolone produced directly from cholesterol is the precursor molecule for all C_{18}, C_{19} and C_{21} steroids.

7.2 CLASSIFICATION

Steroid hormones are classified into five classes:

1. Androgens
2. Estrogens
3. Progestins
4. Mineralocorticoids
5. Glucocorticoids

7.3 STEROID HORMONE BIOSYNTHESIS

Steroid hormones are synthesised on basis of need. Whenever the body needs a certain process done or needs a certain protein synthesised, the brain releases a signal to produce a certain type of steroid hormone. The signals are transmitted through intermediary peptide hormones.

The production and secretion of steroid hormones are controlled by trophic hormones, which themselves are either proteins or peptides. The following indicates which peptide hormone is responsible for stimulating the synthesis of which steroid hormone:

Peptide Hormone	*Steroid Hormone*
Luteinizing Hormone (LH):	progesterone and testosterone
Adrenocorticotropic hormone (ACTH):	cortisol
Follicle Stimulating Hormone (FSH):	estradiol
Angiotensin II/III:	aldosterone

7.4 STEROIDS OF THE ADRENAL CORTEX

The adrenal cortex is responsible for production of 3 major classes of steroid hormones: glucocorticoids, which regulate carbohydrate metabolism; mineralocorticoids, which regulate the body levels of sodium and potassium; and androgens, whose actions are similar to that of steroids produced by the male gonads. Adrenal insufficiency is known as Addison disease, and in the absence of steroid hormone replacement therapy, can rapidly cause death (in 1–2 weeks).

The adrenal cortex is composed of 3 main tissue regions: zona glomerulosa, zona fasciculata, and zona reticularis. Although the pathway to pregnenolone synthesis is the same in all zones of the cortex, the zones are histologically and enzymatically distinct, with the exact steroid hormone product dependent on the enzymes present in the cells of each zone.

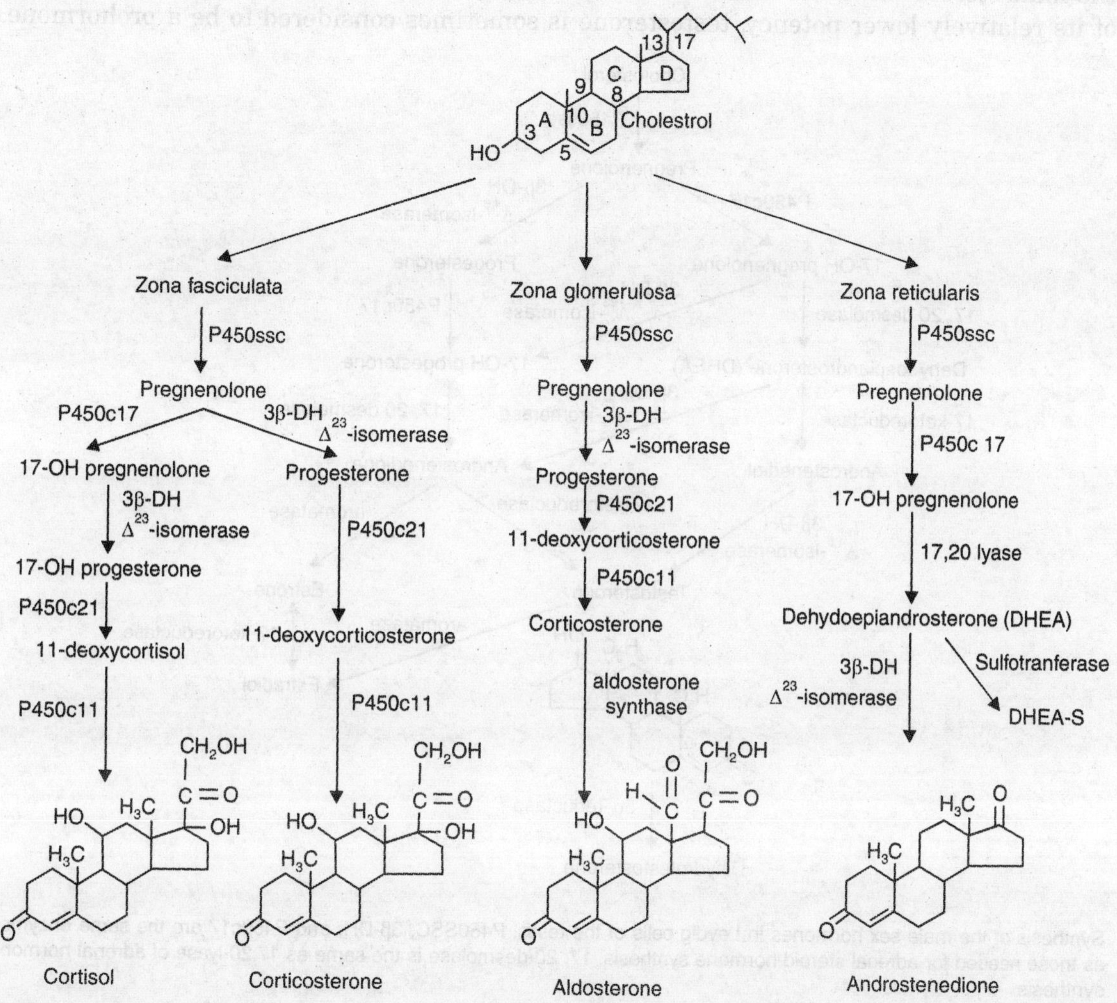

Synthesis of the various adrenal steroid hormones from cholesterol. Only the terminal hormone structures are included. 3β-DH is 3β-dehydrogenase, P450c11 is 11β-hydroxylase, P450c17 is 17α-hydroxylase, P450c21 is 21β-hydroxylase.

7.5 GONADAL STEROID HORMONES

Although many steroids are produced by the testes and the ovaries, the two most important are testosterone and estradiol. These compounds are under tight biosynthetic control. They regulate the secretion of follicle stimulating hormone (FSH) and luteinising hormone (LH) by the pituitary and gonadotropin releasing hormone (GnRH) by the hypothalamus.

The biosynthetic pathway to sex hormones in male and female gonadal tissue includes the production of the androgens. Testes and ovaries contain an additional enzyme that enables androgens to be converted to testosterone. In a number of target tissues, testosterone can be converted to dihydrotestosterone (DHT). DHT is the most potent of the male steroid hormones with an activity that is 10 times that of testosterone. Because of its relatively lower potency, testosterone is sometimes considered to be a prohormone.

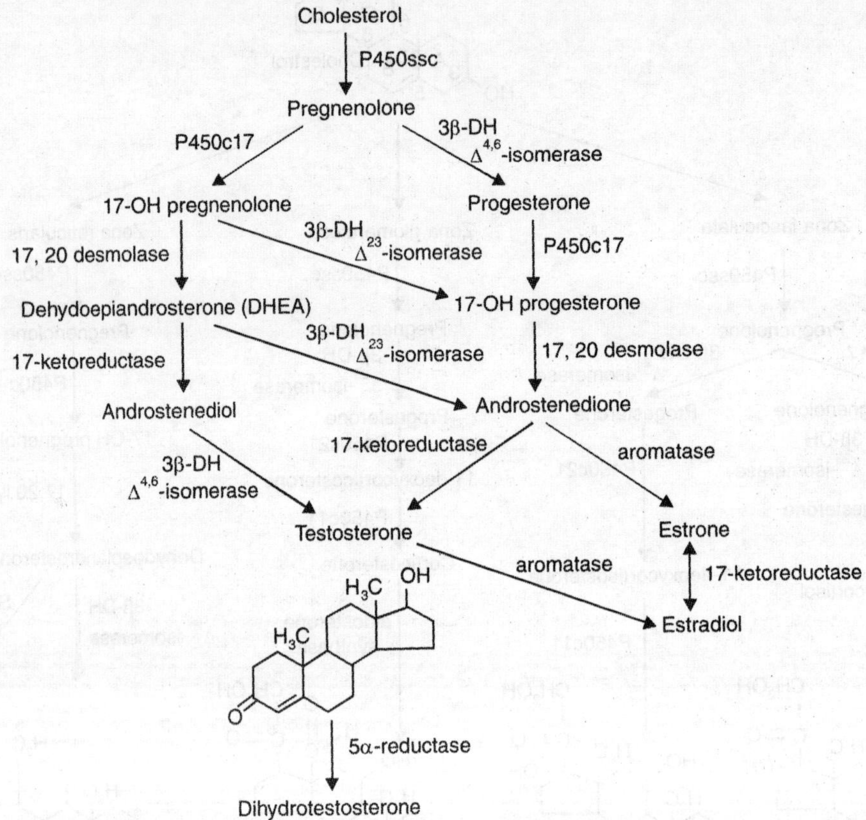

Synthesis of the male sex hormones in Leydig cells of the testis. P450SSC, 3β-DH, and P450c17 are the same enzymes as those needed for adrenal steroid hormone synthesis. 17, 20-desmolase is the same as 17,20-lyase of adrenal hormone synthesis.

Synthesis of the major female sex hormones in the ovary. Synthesis of testosterone and androstenedione from cholesterol occurs by the same pathways as indicated for synthesis of the male sex hormones.

7.6 FUNCTION

7.6.1 Gonadal Steroid Hormones

Progesterone: a progestin, produced directly from pregnenolone and secreted from the *corpus luteum*, responsible for changes associated with luteral phase of the menstrual cycle, differentiation factor for mammary glands.

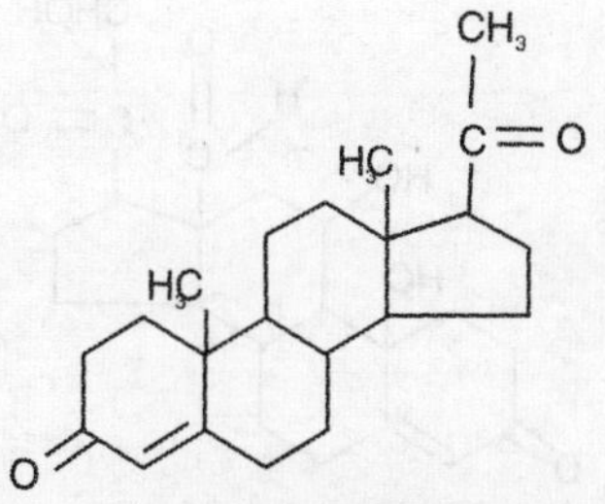

Progesterone

Testosterone: produced from progesterone promotes the normal development of male genital organs and is synthesised from cholesterol in the testes. It also promotes secondary male sexual characteristics such as deep voice, facial and body hair.

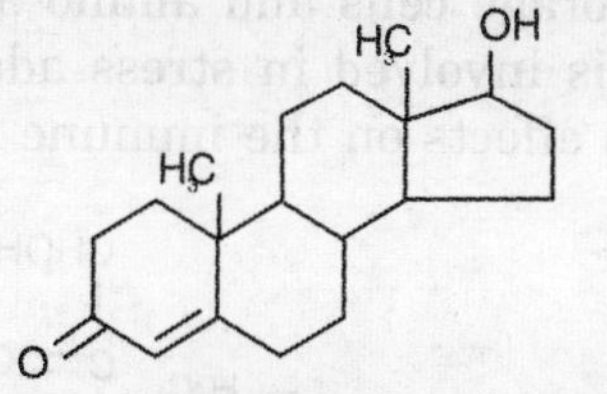

Testosterone

Estradiol: an estrogen, principal female sex hormone, produced in the ovary, responsible for secondary female sex characteristics.

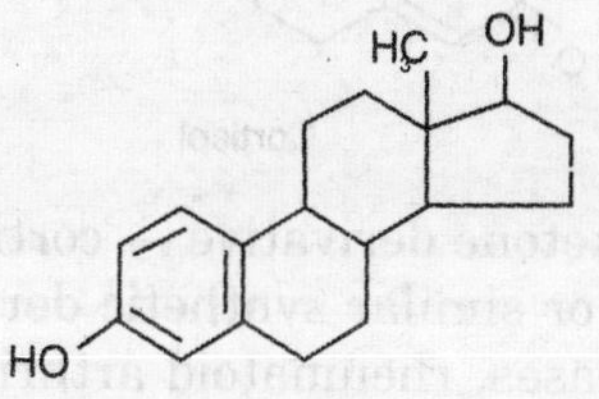

Estradiol

7.6.2 Adrenocorticoid Hormones

The adrenocorticoid hormones are products of the adrenal glands, "adrenal" means adjacent to the renal (kidney).

The most important mineralocorticoid is aldosterone, produced from progesterone in the *zona glomerulosa* of adrenal cortex — regulates the reabsorption of sodium and chloride ions in the kidney tubules and increases the loss of potassium ions. Aldosterone is secreted when blood sodium ion levels are too low to cause the kidney to retain sodium ions. If sodium levels are elevated, aldosterone is not secreted, so that some sodium will be lost in the urine. Aldosterone also controls swelling in the tissues.

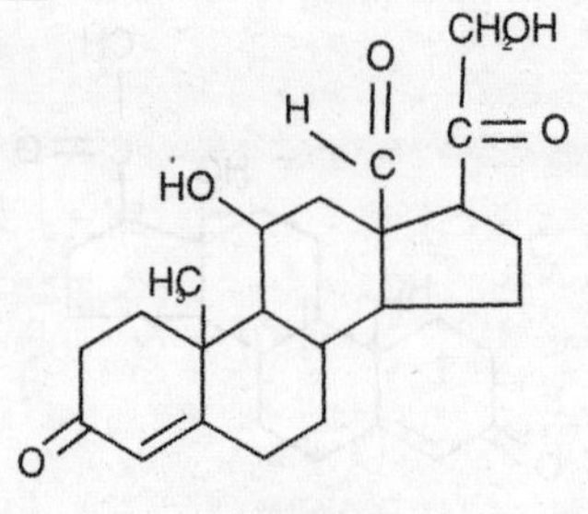

Aldosterone

Cortisol, the most important glucocorticoid in the humans, synthesised from progesterone in the *zona fasciculata* of the adrenal cortex, has the function of increasing glucose and glycogen concentrations in the body. These reactions are completed in the liver by taking fatty acids from lipid storage cells and amino acids from body proteins to make glucose and glycogen. Cortisol is involved in stress adaptation, elevates blood pressure and Na^+ uptake, and numerous effects on the immune system.

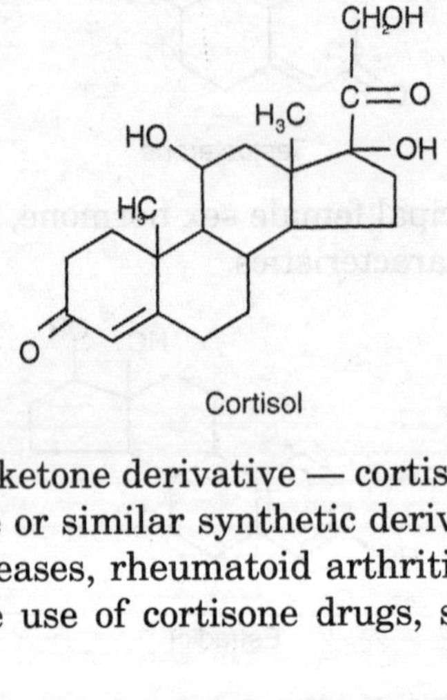

Cortisol

In addition, cortisol and its ketone derivative — cortisone, have the ability to alleviate inflammatory effects. Cortisone or similar synthetic derivatives such as prednisolone are used to treat inflammatory diseases, rheumatoid arthritis, and bronchial asthma. There are many side effects with the use of cortisone drugs, so their use must be monitored carefully.

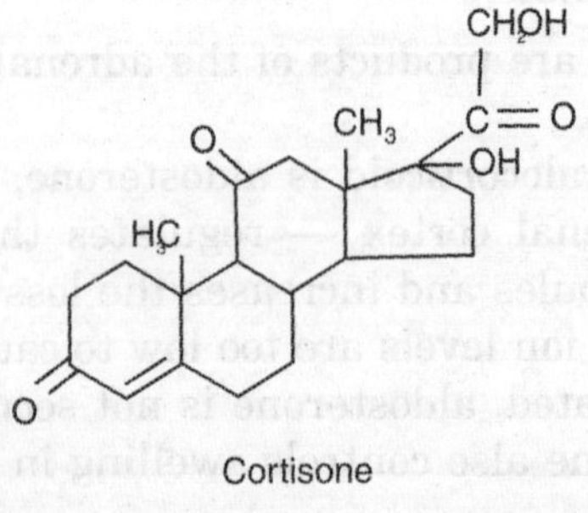

Cortisone

All the steroid hormones exert their action by passing through the plasma membrane and binding to intracellular receptors. The mechanism of action of the thyroid hormones is similar; they interact with intracellular receptors. Both the steroid and thyroid hormone-receptor complexes exert their action by binding to specific nucleotide sequences in the DNA of responsive genes. These DNA sequences are identified as hormone response elements — HREs. The interaction of steroid-receptor complexes with DNA leads to altered rates of transcription of the associated genes.

Part B

Biophysics

1

Scope and Development of Biophysics

Biophysics is truely a multidisciplinary subject linking many areas of physical (physics, chemistry and mathematics) and natural science (Biology, pharmacology and medicine). One of the main thrust of biophysics is quantitative analysis of the functions of biological molecules with reference to its physical and chemical parameters. Although the applications of biophysics is more relevant to biological and medical sciences, its techniques that biophysics relies on is very closer to physics and chemistry.

Many of the new avenues in the field of biological science and applied aspect of biotechnology is due to the advancement made in biophysics. Therefore, biophysics is a trendsetter in progressing in branches of biology.

1.1 DEVELOPMENT OF BIOPHYSICS

Many of the diversified subjects in biology such as molecular biology, genetics, biochemistry and microbiology rely on the physical techniques to understand the functions of biological molecules, and life processes at the cellular and molecular levels. Until mid 20th century, biology as a whole was only a speculative without much scientific background. Most of the work in biology was perhaps related to classification of organisms and their morphological description. However, several physical scientists have made significant contribution to the biological science to describe functional details of organisms at molecular and atomic level. The most eminent scientists have contributed in the development of biophysics are Max Delbriick, Erwin Schrodinger, George Gamow, Linus pauling etc.

The dramatic change in biology lead to the implications of physico-chemical methods for understanding of the complex arrangement of living matter. Several biological problems have been addressed at the cellular and molecular levels, due to progress made in physical science and computer modelling which was thus effectively applied to biological systems. Some of the outstanding contribution of physical science in the development of biophysics include direct application of electron microscopy, X-ray diffraction, spectroscopy, magnetic resonance and X-ray diffraction.

1.2 SCOPE

The immense potential of biophysics is noticed in several diversified fields like pure biology, medicine, agriculture etc. In the following headlines scope of biophysics have been highlighted:

1.2.1 Understanding Molecular Spectrum

The contribution of X-ray diffraction, spectroscopy and electron microscopy have made significant progress in understanding structural and organisation of living organisms. The interactive nature of molecules during developmental process was established by applying various advanced techniques.

1.2.2 Microbiology

Earlier work in microbiology clearly shows that classification and speculative description were only the viable options. However, contribution of electron microscopy and X-ray diffraction made impetus in understanding structure of bacterial cell wall, membranes, nature of flagella and cellular constituents. The reproductive nature was possible to understand only by using high powered resolved microscopes.

1.2.3 Molecular Biology

The progress made in the development in molecular chemistry particularly by hydrodynamical methods, trace and enzymes-labelled technique and contributing to bioenergetics have made significant progress in biophysical science.

The contribution of NMR technique in understanding the solutions facilitated in determining the three-dimensional structures of certain important biomolecules like nucleic acid and proteins. Their small molecular sign have been determined with the advancement of many of these techniques. Remarkable advancement made in molecular biology in recent years has also contributed in bringing biophysics to core stage.

Discovery of unique techniques like gene amplification by polymerase chain reaction (PCR), protein engineering by site-directed mutagenesis, allowed greater sophistication and broader handling of methods to biologist, to engineer or manipulate macromolecules.

The molecular tinkering by computational modelling is mainly due to understanding of three-dimensional structures by X-ray diffraction and NMR methods. These biophysical techniques and computational modelling played a significant role in the development progress of biological science.

1.2.4 Medicine

Contribution of many biophysical techniques, like X-ray, NMR, radioisotope ultrasound, echoencephalograph (EEG), electrocardiogram (ECG) and CT/CAT scan have played immense role in diagnosis of structural disorder or deformities in humans.

X-ray imaging of fractured bones, scull can facilitate diagnosis. Tumor diagnosis by NMR studies often offer ideal clues on cancer treatment. Radiation has been used in the treatment of cancer long time. Therefore, radioisotopic technique is very important in these processes.

1.2.5 Agriculture

Improvement of productivity by mutation treatment requires biophysical technique like radiation. Tracing of disease resistant proteins by biophysical technique is one of the primary focus in the field of agriculture. Tracer technique in elucidating enzymatic proteins involved in primary and secondary metabolic pathway has been found to have important role in metabolic engineering. Understanding events in photosynthesis, redox potential, mechanism of electron transport chains, and some of the key contributions of biophysical science in agriculture have greater impact on productivity of crop plants.

2

pH and Buffer Concepts

Most biological processes in the cell take place in a water based environment. Pure water molecules have a tendency to undergo reversible ionisation to yield a proton (H^+) and a hydroxide ion (OH^-). Hence water conducts electricity.

$$\underset{\text{Acid}(1)}{H_2O} + \underset{\text{Base}(2)}{H_2O} \rightleftharpoons \underset{\text{Acid}(2)}{H_3O^+} + \underset{\text{Base}(1)}{OH^-}$$

Water is amphiprotic. It can either donate a proton or accept a proton. The degree of ionisation of water is very small at equilibrium. At 25° C, it is 1×10^{-7} M *i.e.*, for $\frac{10^7}{55.5}$ molecules of water (500 million) only one molecule of water is ionised. The equilibrium constant of water is given by K_{Eq}

$$K_{Eq} = \frac{[H_3O^+][OH^-]}{[H_2O]}$$

However, the concentration of $[H_3O^+]$ and $[OH^-]$ is very low (1×10^{-7}) when compared to the concentration of water (55.5 M) and is essentially constant.

$$K_{Eq} = \frac{[H_3O^+][OH^-]}{55.5\text{ M}}$$

$$(K_{Eq})(55.5\text{ M}) = [H_3O^+][OH^-] = K_w$$

At 25° C, K_W designates the product (K_{Eq}) (55.5 M) and is called the Ion Product of Water. The value of K_{Eq} determined by electrical conductivity measurements of pure water is 1.8×10^{-16} M at 25° C.

$$K_W = (55.5\text{M})(K_{Eq}) = [H_3O^+][OH^-]$$

$$= (55.5\text{M})(1.8 \times 10^{-16}\text{ M})$$

$$= 1.0 \times 10^{-14}\ M^2$$

$$\therefore \quad [H_3O^+][OH^-] = 1.0 \times 10^{-14}\ M^2$$

When there are exactly equal concentrations of both $[H_3O^+]$ and $[OH^-]$, as in pure water, the solution is said to be neutral, and hence the concentration of $[H^+]$ and $[OH^-]$ can be calculated from the Ion Product of water.

$$K_w = [H^+][OH^-] = [H^+]^2 \text{ since } [OH^-] = [H^+]$$

$$[H^+] = \sqrt{K_w} = \sqrt{1.0 \times 10^{-14}\ M^2}$$

$$\therefore \quad [H^+] = 1.0 \times 10^{-7}\ M = [OH^-]$$

Acidic and basic molecules, when dissolved in water in a biological cell or test tube, react with either H^+ or OH^- and shift the equilibrium.

pH [Hydrogen-ion exponent]

For many purposes, especially when dealing with small concentrations, it is cumbersome to express concentrations of hydrogen and hydroxyl ions in terms of moles per cubic decimeter (M). A very convenient method was proposed by Danish Biochemist Soren Sorenson in 1909. He introduced the term, hydrogen ion exponent, pH.

$$[H^+] = 10^{-pH} \quad \text{or} \quad pH = \log_{10} \frac{1}{[H^+]} = -\log_{10}[H^+]$$

The ion product of water, K_W, is the basis for the pH scale. It is a convenient way of designating the concentration of H^+ and OH^- in any aqueous solution in the range between 1.0 M $[H^+]$ and 1.0 M $[OH^-]$. This is expressed as a positive number between 0 and 14.

A solution with a $[H^+]$ of 1 mole/dm^3, has a pH of 0.

$$pH - \log_{10}[H^+] = -\log_{10} 1.0 = 0 \quad \text{or} \quad [H^+] = 10^{\circ} = 0$$

A solution with a $[OH^-]$ of 1 mole/dm^3, has a pH of 14.

$$[H^+] = \frac{K_w}{[OH^-]} = \frac{10^{-14}}{1} = 10^{-14}$$

$$pH = -\log_{10}[H^+]$$

$$= -\log_{10} 10^{-14}$$

$$= -(-14 \log_{10} 10)$$

$$= 14 \log_{10} 10 = 14 \times 1 = +14$$

For a neutral solution at 25° C, the concentration of hydrogen is 1×10^{-7}.

$$pH = -\log_{10}[H^+] = -\log_{10} 10^{-7}$$

$$= -(-7\log_{10} 10) = +7 \times 1 = +7$$

Measurement of pH is one of the most important and frequently used procedures in biochemical laboratories. The pH affects the structure and activity of biological macro-molecules, for example, the catalytic activity of enzymes is strongly dependent on pH.

The logarithmic or exponential method has also been used for expressing dissociation constants of weak acids.

$$K_a = 10^{-pK_a} \quad \text{or} \quad pk_a = -\log_{10} K_a$$

The ionisation of acetic acid is represented as:

$$\underset{\text{Acid(1)}}{CH_3COOH} + \underset{\text{Base(2)}}{H_2O} \rightleftharpoons \underset{\text{Acid(2)}}{H_3O^+} + \underset{\text{Base(1)}}{CH_3COO^-}$$

$$K_a = \frac{[H_3O^+][CH_3COO^-]}{[CH_3COOH]} = 1.8 \times 10^{-5}\ M$$

The dissociation of weak base such as ammonia is represented as:

$$\underset{\text{Base(1)}}{NH_3} + \underset{\text{Acid(2)}}{H_2O} \rightleftharpoons \underset{\text{Acid(1)}}{NH_4^+} + \underset{\text{Base(2)}}{OH^-}$$

$$K_b = \frac{[NH_4^+][OH^-]}{[NH_3]} = 1.8 \times 10^{-5}\ M$$

In most dilute aqueous solutions, the quantity of solute is very small compared to the quantity of water. The solvent water is treated like a pure liquid and its concentration does not appear in the equilibrium constant.

Hydrochloric acid is a strong acid and ionises completely

$$\underset{\text{Acid(1)}}{HCl} + \underset{\text{Base(2)}}{H_2O} \longrightarrow \underset{\text{Acid(2)}}{H_3O^+} + \underset{\text{Base(1)}}{Cl^-}$$

The tendency for Cl^- to accept a proton from H_3O^+ is very limited compared to the tendency for HCl to donate a proton to H_2O.

Determination of pK_a values of weak Acids

Titration is used to determine the amount of an acid in a given solution. A measured volume of the acid is (acetic acid) titrated with a solution of strong base (NaOH) of known concentration. The NaOH is added in small increments until the acid is neutralised as determined by pH meter. The concentration of the acid is calculated from the volume and concentration of NaOH added.

A plot of pH against the amount of NaOH added (a titration curve) reveals the pK_a of the weak acid. Consider the titration of a 0.1 M solution of acetic acid with 0.1 M NaOH at 25° C. Two reversible equilibria are involved in the process.

$$H_2O \rightleftharpoons H_3O^+ + OH^- \qquad K_w = [H_3O^+][OH^-] = 1 \times 10^{-14}\ M^2$$

$$HAc \rightleftharpoons H^+ + Ac^- \qquad K_a = \frac{[H^+][Ac^-]}{HAc} = 1.74 \times 10^{-5}\ M$$

At the beginning of the titration, before any NaOH is added, the acetic acid is already slightly ionised (1.3×10^{-3}) and can be calculated from the K_a value.

$$\frac{[H^+][Ac^-]}{[HAc]} = 1.74 \times 10^{-5} \qquad [HAc] = 1 \times 10^{-1}\ M$$

$$[H^+][Ac^-] = 1.74 \times 10^{-5} \times 1 \times 10^{-1} = 1.74 \times 10^{-6}$$

Since $[H^+] = [Ac^-]$

$$[H^+]^2 = 1.74 \times 10^{-6}$$

$$[H^+] = 0.0013 = 1.3 \times 10^{-3}\ M$$

$$pH = -\log 1.3 \times 10^{-3} = -(-3)\log 1.3$$

$$= 3 - 0.1139$$
$$= 2.8861$$

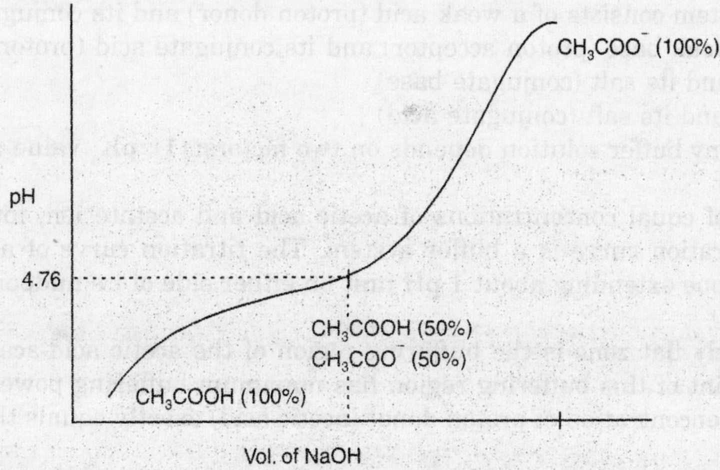

As NaOH is gradually introduced, the added OH^- combines with the free H^+ in the solution to form H_2O. As free H^+ is removed, HAc dissociates further to satisfy its own equilibrium constant.

The net result as the titration proceeds is that more and more HAc ionizes, forming Ac^-, as the NaOH is added. At the mid point of the titration, at which exactly 0.5 equivalent of NaOH has been added, one-half of the original acetic acid has undergone dissociation, so that the concentration of the proton donor (HAc), now equals that of the proton acceptor, (Ac^-).

At this mid-point, there is a very important relationship. The pH of the equimolar solution of acetic acid and acetate is exactly equal to the pK_a of acetic acid (pK_a = 4.76).

As the titration is continued by adding further increments of NaOH, the remaining undissociated acetic acid is gradually converted into acetate. The end point of the titration occurs at about pH 7.0. All the HAc has lost its protons to OH^- to form H_2O and Ac^-.

Acid		pK_a			pK_a
Formic acid		3.75	Ammonia		9.24
Acetic acid		4.76	Tris (Hydroxymethyl		
Citric acid	pK_{a1}	3.13	aminomethane)		8.08
	pK_{a2}	4.76	Glycine	pK_1	2.34
	pK_{a3}	6.40		pK_{a2}	9.60
Phosphoric acid	pK_{a1}	2.14			
	pK_{a2}	7.21			
	pK_{a3}	12.35			
Carbonic acid	pK_{a1}	6.37			
	pK_{a2}	10.33			
Boric acid		9.24			

The most important point about the titration curve of a weak acid is that it shows that a weak acid and its anion– a conjugate acid-base pair- can act as a buffer.

Buffers

Buffers are aqueous systems that tend to resist changes in pH when small amounts of $[H^+]$ or $[OH^-]$ (acid or base) are added.

A buffer system consists of a weak acid (proton donor) and its conjugate base (proton acceptor) or a weak base (proton acceptor) and its conjugate acid (proton donor).

Weak acid and its salt (conjugate base)

Weak base and its salt (conjugate acid)

The pH of any buffer solution depends on two factors: (1) pK_a value and (2) the ratio of salt to acids.

A mixture of equal concentrations of acetic acid and acetate ion, found at the mid-point of the titration curve is a buffer system. The titration curve of acetic acid has a relatively flat zone extending about 1 pH unit on either side of its midpoint pH of 4.76 or pK_a value (4.76).

The relatively flat zone is the buffering region of the acetic acid-acetate buffer pair, and the mid-point of this buffering region has maximum buffering power or capacity. At this point, the concentration of proton donor (acetic acid) exactly equals that of the proton acceptor (Acetate). In this region, the pH changes least on addition of H^+ or OH^-. The pH at this point in the titration curve of acetic acid is equal to its pK_a.

The pH of the acetate buffer system changes slightly when a small amount of H^+ or OH^- is added. But this change is very small compared with the pH change that would result if the same amount of H^+ or OH^- were added to pure water or to a solution of the salt of a strong acid and strong base such as NaCl, which has no buffering power.

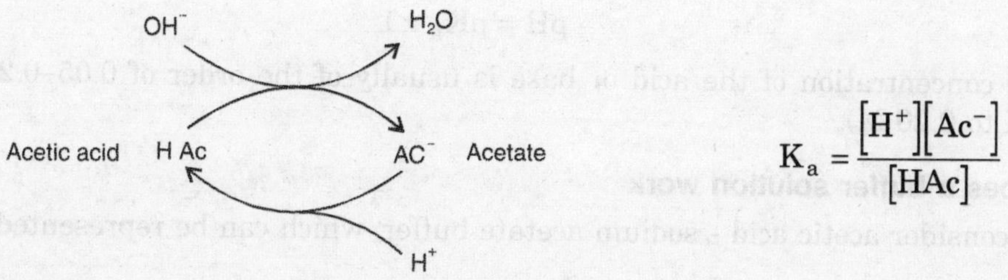

$$K_a = \frac{[H^+][Ac^-]}{[HAc]}$$

The relationship between pH, pK_a and the molar concentrations of the two components is given by Henderson-Hasselbalch Equation.

$$K_a = \frac{[H^+][A^-]}{[HA]}$$

$$[H^+] = K_a = \frac{[HA]}{[A^-]}$$ Take negative logarithm on both sides

$$-\log[H^+] = -\log K_a - \log\frac{[HA]}{[A^-]}$$

$$pH = pK_a - \log\frac{[HA]}{[A^-]}$$ Now invert $-\log\frac{[HA]}{[A^-]}$

$$pH = pK_a + \log\frac{[A^-]}{[HA]}$$

$$pH = pK_a + \log\frac{[\text{proton acceptor}]}{[\text{proton donor}]}$$

At the mid-point of a titration, $[HA] = [A^-]$

$\therefore$ $pH = pK_a + \log 1.0$

$= pK_a + 0$

$pH = pK_a$

Henderson-Hasselbalch equation also makes it possible to

(1) calculate pH, given pK_a and the molar ratio of proton donor and acceptor
(2) calculate the molar ratio of proton donor and acceptor, given pK_a and pH
(3) calculate pK_a, given pH and the molar ratio of proton donor and acceptor

The buffering capacity is maintained within the range 1 acid: 10 salt and 10 acid: 1 salt. The approximate pH range of a weak acid buffer is

$$pH = pK_a \pm 1$$

The concentration of the acid or base is usually of the order of 0.05–0.20 mol dm^{-3} (0.05 M to 0.20 M).

How does a buffer solution work

Let us consider acetic acid - sodium acetate buffer, which can be represented as:

$$CH_3COOH + H_2O \rightleftharpoons H_3O^+ + CH_3COO^-$$

$$K_a = \frac{[H_3O^+][CH_3COO^-]}{[CH_3COOH]} = 1.8 \times 10^{-5}\ M$$

$$[H_3O^+] = K_a \times \frac{[CH_3COOH]}{[CH_3COO^-]} = 1.8 \times 10^{-5} \times \frac{[CH_3COOH]}{[CH_3COO^-]}$$

For 1 M CH_3COOH - 1M CH_3COO Na buffer solution,

$$[H_3O^+] = 1.8 \times 10^{-5} \times \frac{1}{1} = 1.8 \times 10^{-5}$$

$$pH = -\log[H_3O^+] = -\log[1.8 \times 10^{-5}] = +5 - \log 1.8$$

$$= 5 - 0.2553$$

$$= 4.7445$$

Now suppose we add a strong base to neutralise 2% of the acetic acid, this will reduce the concentration of acetic acid to 0.98 M and raise that of acetate ion to 1.02 M.

$$CH_3COOH + OH^- \rightleftharpoons H_2O + CH_3COO^-$$

	CH_3COOH	OH^-	CH_3COO^-
Initial:	1 M		1 M
Add:		0.02 M	
Changes:	–0.02	–0.02	+ 0.02
After neutralisation:	0.98 M	0 M	1.02 M

$$[H_3O^+] = K_a \times \frac{[CH_3COOH]}{[CH_3COO^-]} = 1.8 \times 10^{-5} \times \frac{0.98}{1.02} = 1.729 \times 10^{-5}$$

$$pH = -\log 1.729 \times 10^{-5} = +5 - \log 1.729 = 5 - 0.2377 = \mathbf{4.7623}$$

The [H+] and pH of the buffer solution is hardly affected.

However, if 0.02 M acid is added to the buffer solution

$$CH_3COO^- + H_3O^+ \rightleftharpoons H_2O + CH_3COOH$$

	CH_3COO^-	H_3O^+	CH_3COOH
Initial:	1.0 M		1.0 M
Add:		0.02 M	
Changes:	–0.02	–0.02	+ 0.02
After neutralisation:	0.98 M	0 M	1.02 M

$$\left[H_3O^+\right] = K_a \times \frac{[CH_3COOH]}{\left[CH_3COO^-\right]} = 1.8 \times 10^{-5} \times \frac{1.02}{0.98} = 1.8735 \times 10^{-5}$$

$$pH = -\log 1.8735 \times 10^{-5} = +5 - \log 1.8735 = 5 - 0.2727 = \mathbf{4.7273}$$

Again the $\left[H_3O^+\right]$ and pH are hardly affected.

$$pH = pK_a + \log \frac{\left[CH_3COO^-\right]}{[CH_3COOH]} = 4.7447 + \log 1.041 = 4.7447 + 0.0174 = 4.7621$$

$$pH = pK_a + \log \frac{\left[CH_3COO^-\right]}{[CH_3COOH]} = 4.7447 + \log 0.9608 = 4.7447 + \bar{1}.982$$

$$= 4.7447 - 0.0173$$

$$= 4.7274$$

However, if the buffer was not in the solution, the pH would have been 12.3 or 1.7, when 0.02M NaOH or 0.02M HCl was added, respectively.

$$pH = -\log\left[H^+\right] = -\log 2 \times 10^{-2} = +2 - \log 2 = 2 - 0.03 = 1.70$$

$$pOH = -\log\left[OH^-\right] = -\log 2 \times 10^{-2} = +2\log 2 = 2 - 0.30 = 1.70$$

$$\therefore\ pH = 14 - pOH = 14 - 1.70 = 12.30.$$

3

Chemical Bonding

3.1 STRUCTURE OF ATOM

All atoms comprise of a nucleus in the centre surrounded by electrons. There are one or more electrons revolving around the nucleus. The nucleus contains protons and neutrons alongwith. The protons exhibit positive charge whereas neutrons are electrically neutral. The mass of neutron is equivalent to proton in the nucleus. Electrons on the contrary are negative and are very light in weight *i.e.*, about 1/1836 the weight of proton. The neutral nature of atom is due to the negative charges of electrons which counter balance the positive charges of protons.

The electrons are similar to light in having characters like particle and a wave. Electrons move around nucleus in a discrete volume called shell with a definite energy. Electrons in each shell are placed at a particular distance from the nucleus. The electrons in a shell closer to the nucleus exhibit lowest energy. Shells are designated as K, L, M, N, O, P and Q with a corresponding energy 1, 2, ... 7. The electron occupies a lowest energy state. It is stated as, n = 1, the next n = 2 and next n = 3 and so on. The lowest energy state is in n = 1 — K shell. Each shell can accommodate $2n^2$ electrons *i.e.*, K shell contains two electrons. The second shell can accommodate eight electrons and the third state or third shell contains 18 etc. The principal quantum number is the number of shells represented by the symbol n.

The electrons are found possibly in specific regions within each shell. These regions exhibit particular shape and are known as orbitals. Based on their shape there are four different types of orbitals and are designated as s, p, d and f orbitals. The s orbitals are spherical in shape. The p orbitals exhibit dumb-bell-shape. Similarly d orbitals are double dumb-bell-shaped and finally f orbitals possess various shapes.

3.2 CHEMICAL BOND

Most of the atoms tend to aggregate into molecules which are electrically neutral. Chemical bond is simply an attraction between the atoms within a molecule. Several kinds of chemical bonds between atoms and molecules are important in biological systems. The strongest and most stable are the covalent bonds between atoms. Non covalent bonds such as hydrogen bonds, ionic and Van der Waals interactions are weaker bonds but they help to stabilise biological structures like proteins and nucleic acids, as well as other biomolecular complexes. Non covalent bonds are indispensable for maintaining the architecture of many biological molecules and membranes.

3.2.1 Covalent Bond

A covalent bond is formed between two atoms when they share an electron pair between them *i.e.*, atoms are held together because electrons in their outermost shells move in orbitals that are shared by both atoms. As a result, these electrons are common to both the nuclei and does not belong to any one atom. For example, two hydrogen atoms, containing one electron can attain a stable configuration. This could be possible because they are sharing the electron between them. In addition, the 's' orbitals of both the atoms overlap the pair of electrons and shift to a new orbital where they are electrostatically attracted towards both the nuclei of the hydrogen atom. The combination of two atomic orbitals always leads to molecular orbitals, and the bond formation between these two atoms is known as covalent bond.

Most of the molecules in living system contain only six different atoms; these are carbon, hydrogen, oxygen, nitrogen, phosphorus and sulphur. The outer electron shell of each atom exhibit characteristic number of electrons.

$$\overset{\bullet}{\text{H}},\quad \bullet\overset{\bullet}{\underset{\bullet}{\text{C}}}\bullet,\quad \bullet\overset{\bullet\bullet}{\underset{\bullet}{\text{N}}}\bullet,\quad \bullet\overset{\bullet\bullet}{\underset{\bullet}{\text{P}}}\bullet,\quad \bullet\overset{\bullet\bullet}{\underset{\bullet\bullet}{\text{O}}}\bullet,\quad \bullet\overset{\bullet\bullet}{\underset{\bullet\bullet}{\text{S}}}\bullet$$

These atoms can readily form covalent bonds with other. Their electron orbitals can overlap into molecular orbitals. In all these atoms only the outermost unpaired electrons participate in covalent bonds. Each type of atom in biological system forms a characteristic number of covalent bonds with other atoms. A hydrogen atom having only one electron forms only one bond. A carbon atom having four electrons in its outermost shell can form four bonds. For example, in methane four covalent bonds can be seen.

```
      H
      |
H  —  C  —  H
      |
      H
```

Oxygen and sulphur contain six electrons in their outermost shells and atom of these elements usually forms only two covalent bonds as in the case of molecular oxygen and hydrogen sulphide. On the contrary, sulphur atom can form six covalent bonds as in sulphuric acid.

While constructing or breaking of covalent bonds large amount of energy changes can take place. In this contrast if a chemical bond requires large amount of energy to break them are considered stronger and more stable bond when compared to the bond that requires less energy to break them. In order to break covalent bond in C-C of gas ethane, requires 83 kCal/mol.

3.2.2 Weak Bonds Stabilize the Structure of Biomolecules

The strongest forces that hold the atoms of a molecule together are covalent forces. The typical amount of energy released in covalent bond formation ranges from 50 to 200 kCal/mol. However, it is not the monopoly of the covalent bonds responsible for maintaining the structure of large molecules. The stability of three-dimensional structure of large molecules often stems from weaker bonds. The amount of energy released in the formation of these weak bonds is only 1 to 5 kcal/mol when compared to the status of covalent bond of 50 to 200 kcal/mol. Many of these weak bonds can act together to produce very stable structure. In certain cases, the weak bonds that hold molecules together can break from one another. For example, when two protein molecules bound together by four weak bonds and if two of the four bonds break, the protein molecules will still bind together by the other two bonds. Even if two of the bonds are broken, the remaining two bonds will facilitate the reformation of the broken ones. This is one of the reason why certain molecules can maintain stable associations with other molecules if enough weak bonds work together (Fig. 3.1). Four main types of weak bonds which play a crucial role in maintaining biological structures are the hydrogen bond, the ionic bond, the Van der Waals interaction and the hydrophobic bond or interaction.

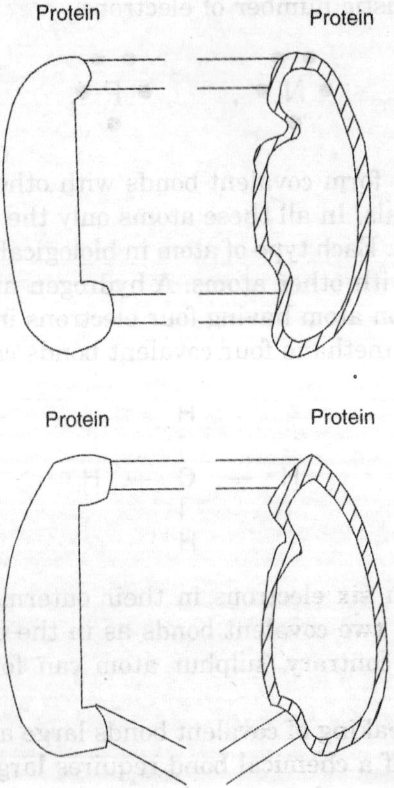

Fig. 3.1 Role of weak bonds in stabilizing two protein molecules

3.2.2.1 The Hydrogen Bond

A hydrogen bond is a weak association between an electronegative atom (X) and a hydrogen atom that is linked to another atom (Y) by covalent bond. At this juncture hydrogen atom is closer to the 'X' than to the Y atom.

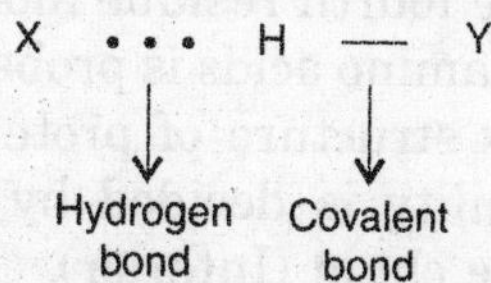

The covalent bond between the donor (Y) and the hydrogen atom must be dipolar, and the outer shell of the acceptor atom must have non bonding electrons that attract the σ + ve charge of the hydrogen atom. The classic example is the hydrogen bond in water. The hydrogen atom in one molecule is attracted to a pair of electrons in the outer shell of an adjacent oxygen atom. The strength of this bond in water is 5 kCal/mol which is weaker than a covalent bond (H-O bond). Hydrogen bonding is the interaction arising between covalently bonded hydrogen atom of a proton donor group (A-H) and a region of high electron density of a proton acceptor group B. The letter A is an electronegative atom of Oxygen (O), Nitrogen (N), Sulphur (S).

Hydrogen bonding is responsible for the reduction of free molecules and also enhances the average molecular weight. The exothermic and endothermic process can be seen during the formation and breaking of hydrogen bond respectively. Certain molecules (water, alcohol, carboxylic acids, primary and secondary amines) with both donor and acceptor group can form hydrogen bonding whereas molecules with neither donor nor acceptor group (saturated hydrocarbons) cannot participate in hydrogen bonding. Hydrogen bonding is of two types, one is intermolecular and other intramolecular. The intermolecular hydrogen bonding tends to aggregate molecules under the influence of shape of the molecule. The molecular weight of a structure is increased due to intermolecular hydrogen bonding. In certain molecules like water, carboxylic acids, primary and secondary amines, intramolecular hydrogen bonding takes place. The bonding takes place only when the proton donor and one proton acceptor sites on the same molecule is in a favourable spatial configuration. This type of bonding also occurs due to the cis-configuration of the molecules.

3.2.2.1.1 H Bonding in biological system: One of the most classical example for hydrogen bonding can be seen in water. Nearly ninety per cent of living cell contains water; thus all the biomolecules occur in aqueous environment. The biomolecules which are nearer to water molecule have a greater impact in the structure and conformation of biopolymers. In all living systems these biopolymers enter into intermolecular hydrogen bonding among themselves and with adjoining water molecules and almost all polar molecules are interlinked by a network of hydrogen bonded water molecules in a living system. The most predominant and extensive hydrogen bonding within a cell is responsible for proper organisation of non-polar lipid molecules.

3.2.2.1.2 Hydrogen bonding in proteins: Both intermolecular and intramolecular hydrogen bonding occurs in the peptide molecule. In the pleated sheet arrangement, hydrogen bonding occurs between the polypeptides, whereas intramolecular hydrogen bonding takes place in helical conformation of peptides. The hydrogen bonding is responsible for the secondary structure of protein where it takes place between the carboxyl (C = O) and amino group (N – H) of the peptide bond.

In the helical conformation of peptide, every imino group forms hydrogen bond (2.74 Å long) with the carboxyl group of the fourth residue along the peptide chain. Formation of hydrogen bond in the side chains of amino acids is probably a guiding force in transforming α-helix into tertiary or quaternary structure of protein. The extent of affinity of water towards proteins and water solubility is decided by the hydrogen bond in polar side chains (exterior) and non-polar side chain (Interior).

3.2.2.1.3 Hydrogen bonding in carbohydrate: Hydrogen bond is present in disaccharides and polysaccharides and its bond distance varies from 2.65 to 3.03Å. Sucrose contains two intramolecular hydrogen bonds. In the polysaccharides, extensive hydrogen bonding is responsible for the aggregation of cellulose fibrils into microfibrils. Cellulose chain has a zig-zag conformation with successive glucose residues rotated 180° with respect to each other permitting a hydrogen bond between the –OH group at carbon 3 of one glucose residue with ring oxygen of next glucose residue. The stability and organisation of primary cell wall in plant system is due to multiple hydrogen bonds.

3.2.2.1.4 Hydrogen bonding in nucleic acids: Hydrogen bonding provides stabilizing force to the DNA molecule. Association of two strands of double helical DNA is due to the extensive hydrogen bonding. Bonding takes place between the purines and the pyrimidines. Adenine is linked to thymine/uracil by two hydrogen bonds while guanine pairs cytosine by three hydrogen bonds. The nature of hydrogen bonding in hydrophobic interactions and base stalking play an important role in providing stability to the DNA molecule in an aqueous environment. The multiple hydrogen bonds are broken when the DNA is heated. The atoms are farther apart from one another in a weak bond than in a strong bond. Cooperative hydrogen bonding is a characteristic of both protein and nucleic acid molecules which may contain dozens, hundreds or even thousands of cooperative hydrogen bonds. Cooperativity of hydrogen bonding in biological macromolecules depends on the relative positions of the functional groups capable of forming hydrogen bonds.

3.2.2 Nature of Structural Bonds Between Proteins

Four main types of interactions take place between protein molecules (Intramolecular interactions) in the cell. They are (1) Homopolar cohesive bonds are the type of Van der Waals forces that hold a crystal together. Their bond strength is between 1000 and 2000 calories per mole. In proteins, this type of bond is formed by interlaced methyl groups or other hydrophobic groups of adjacent molecules. Such bonds in proteins are easily broken at slightly higher temperature or heat environment. (2) Heteropolar cohesive bonds, such as dipole to dipole attraction and hydrogen bonds. These result from attraction of neighbouring proteins. These bonds are broken by mild heat, and the bond strength being 5000 calories per mole. (3) Heteropolar valency bonds. These are the coulomb forces. Heteropolar valency bonds are formed by those that form a salt or ester linkage, which are stronger than homopolar cohesive bonds and heteropolar cohesive bonds. Bonds of this nature are not broken by mild heat. (4) Homopolar valency bond, involve formation of a bridge. These bonds like heteropolar valency bonds are also relatively strong bonds of the order of 50,000 calories per mole.

The bonds between phosphatides are significant, since they play an important role in explanations of phenomena as to structure maintained in highly hydrated proteins. In these highly hydrated proteins the above said three bonds are likely to have been broken. Certain ions such as magnesium and calcium help in linking carboxyl groups. Water plays an important role in hydrophilic bonding and also becoming part of the cellular structure through hydrogen bonding (Fig. 3.2).

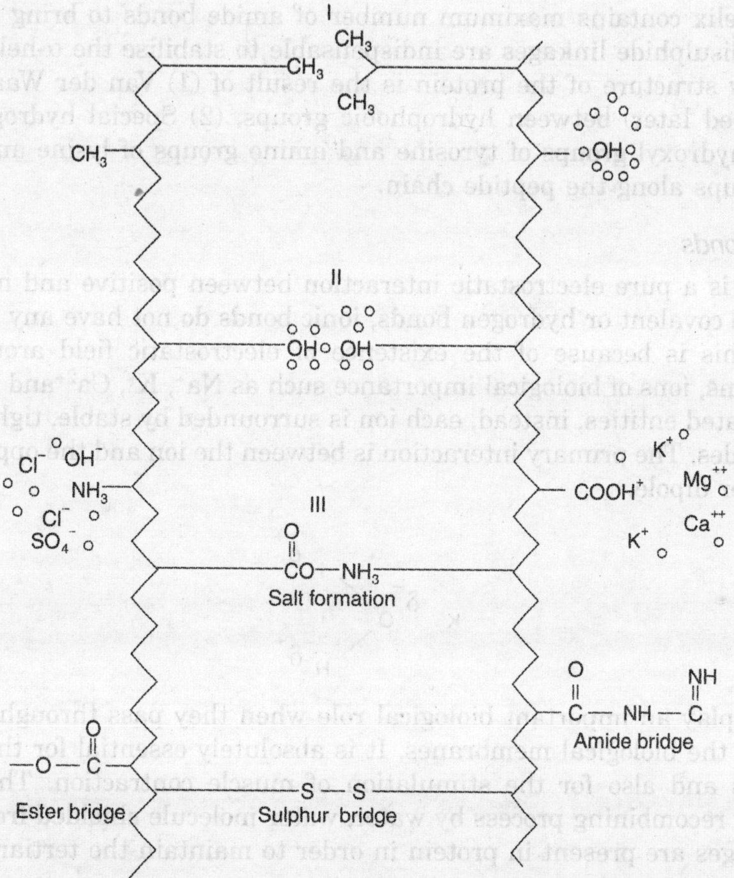

Fig. 3.2 Representation of types of bonds between neighbouring polypeptide chains of cytoplasmic proteins (Inter molecular interactions)

Weak bonds such as homopolar and heteropolar cohesive bonds are important in many biological interactions. For example, they give proper shape to polypeptides and polynucleotides. Strong bonds are unlikely to be broken under physiological conditions. Therefore, they do not provide any structural flexibility to the cytoplasm under changing environmental conditions.

Primary, secondary and tertiary structures occur within the protein molecule. The primary structure of protein depends upon chemical valence bonds. The peptide bonds join amino acid residues in the peptide backbone of a protein, and covalent bonds form cross linkages such as disulphide bonds between cysteine residue.

Secondary structural bonds are due to hydrogen bond formation, for example, the amide (CONH) linkages between C = O and NH^- groups on amino-acid residues of the polypeptide chain. The alpha (α) helix of protein is also due to the result of secondary bonds. The α-helix contains maximum number of amide bonds to bring stability to the molecule. The disulphide linkages are indispensable to stabilise the α-helix in solution.

The tertiary structure of the protein is the result of (1) Van der Waals interactions (will be described later) between hydrophobic groups, (2) Special hydrogen bonds that exist between hydroxyl groups of tyrosine and amino groups of lysine and various electronegative groups along the peptide chain.

3.2.2.2 Ionic Bonds

The ionic bond is a pure electrostatic interaction between positive and negative ions or charges. Unlike covalent or hydrogen bonds, ionic bonds do not have any fixed geometric orientations. This is because of the existence of electrostatic field around an ion. In aqueous solutions, ions of biological importance such as Na^+, K^+, $Ca^{2+}$and Cl^- do not exist as free and isolated entities, instead, each ion is surrounded by stable, tightly bound shell of water molecules. The primary interaction is between the ion and the oppositely charged end of the water dipole.

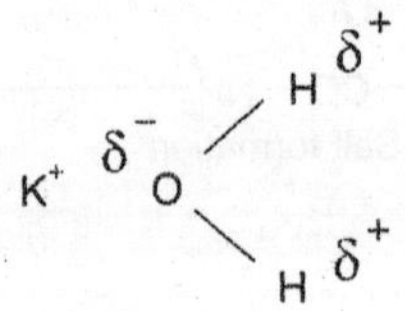

K^+ δ^- O (H δ^+, H δ^+)

These ions play an important biological role when they pass through pores or channels present in the biological membranes. It is absolutely essential for the conduction of nerve impulses and also for the stimulation of muscle contraction. The two ions are prevented from recombining process by water, water molecule shielded from one another. The ionic linkages are present in protein in order to maintain the tertiary structure.

3.2.2.3 Van der Waals Interactions

Van der Waals is a weak attractive force that is produced when two atoms approach one another closely. In an atom any fluctuation in the distribution of its electrons gives rise to transient dipoles. Van der Waals force is operative for small distances and it is being considered as short range force. The proposal on existence of Van der Waals force was made by Dutch physicist Van der Waals. The force would be originated when two dipoles come together within very short distance. There are different types of forces recognised based on the nature of dipoles. One is Keeson force formed due to the interaction of two permanent dipoles *i.e.*, positive end of one dipole is in close proximity to the negative end of the other dipole, which elicites attraction between the molecule; another type is the Debye force. These are the dipole-induced dipole attraction formed due to the polarisation of a non-polar molecule in the electric field of a dipolar molecule leading to an attraction between the two molecules. Van der Waals interactions occur among all types of polar and non-polar molecules, particularly, they are responsible for the attachment among the molecules of nonpolar liquids and solids. The attraction declines rapidly as the distance between these two atoms increases. Thus, force is effective only when atoms are close to one another. However, if atoms get too close together, they get repelled by the negative charges in their outer electron shell.

The energy of the Van der Waals interaction attains about 1kcal/mol. This value is slightly higher than the kinetic energy of molecules at 25° C. In aqueous environment its energy value is 1 to 2 kcal/mol. The interaction between antibody with antigen molecule are the Van der Waals contact. Another Van der Waals interaction is the enzyme and their specific bound substrates.

3.3 PRINCIPLES OF THERMODYNAMICS

Thermodynamics deals with energy relationship of any reaction. Thermodynamics is bound by certain laws and are universal. All physical, chemical and biological processes are bound by these laws. An understanding of the basic concepts of thermodynamics is indispensable for approaching problems in bioenergetics.

The process of thermodynamics are either reversible or irreversible. The reversible process can be seen in chemical reactions, whereas biological systems are unique examples for irreversible process. A system may be closed or open. A *closed* system is the one in which only energy can pass through the system and the system is totally a barrier to the matter. On the contrary in an *open* system, both energy and matter can pass through the system. Examples for closed and open systems are chemical reactions and biological systems respectively.

Cells are living physical and chemical systems that absorb, transduce and effectively utilise energy in maintaining the sustainability of life. Energy is the capacity to do work. Life cannot sustain without expenditure of energy. The principles that govern energy transfer in physical-chemical systems are also applicable to cells, *i.e.*, in general biological system.

Energy is classified into two types — kinetic and potential. The total energy of the molecule is known as internal energy or intrinsic energy. It includes both kinetic and potential energy. All cellular reactions proceed at the backdrop of kinetic energy of the molecules in a cell. Kinetic energy of the molecules is the energy of motion. The kinetic energy of the molecules can be seen from Brownian movement. This movement is attributed to random motion of the molecules colliding with visible particles to which they transfer sufficient energy for movement. Potential energy is either bound or positional. It is energy with high potential for doing work. Conversion of chemical potential energy to other forms of potential energy and to kinetic energy occur continuously in cells. The principles governing the relations between potential and kinetic energy are encased in the laws of thermodynamics.

3.3.1 Law of Thermodynamics

3.3.1.1 First Law of Thermodynamics (Enthalpy)

The first law of thermodynamics is the principle of energy conservation. In any process (physical, chemical or biological) the total energy of the system and its surrounding remains constant. Energy is neither created nor destroyed. However, it may undergo transformation from one form to another, for example, chemical energy may be transformed into thermal, radiant or mechanical energy. This law is associated commonly as Enthalpy (H).

$$\Delta H = \Delta U + P\Delta V$$

where U = internal energy
PV = Pressure, volume
H = Enthalpy
ΔH = change in enthalpy
ΔU = Total energy change
ΔV = change in volume

Endothermic and exothermic reactions can be explained on the basis of enthalpy change. If enthalpy change exceeds zero ($\Delta H > 0$), then energy enters into the system. This reaction is called endothermic whereas if enthalpy change falls below zero, then energy is lost from the system and the reaction process is called exothermic.

If $\Delta H > 0$ the process is endothermic (heat is absorbed)

$\Delta H < 0$ the process is exothermic (heat is released)

Enthalpy can be measured calorimetrically. In biological system, for instance, enthalpy change (ΔH) for the oxidation of glucose is –673,000 cal/mol of glucose at 20° C and 1 atmospheric pressure. As the temperature fluctuates, enthalpy changes are inevitable.

$$C_6H_{12}O_6 + 6O_2 \longrightarrow 6H_2O + 6CO_2$$
$$\Delta H_{293} = 673 \text{ kilo calories}$$

where enthalpy change ΔH is denoted by indication of appropriate absolute temperature in the kelvin scale (0° C = 273.1° K) At (20° C = 273 + 23 = 293° K).

3.3.1.2 The Second Law of Thermodynamics (Entropy)

Entropy is used to denote measure of degree of disorder in any system and is second law of thermodynamics. The entropy of the universe always increases until equilibrium is attained, at this point entropy is maximum under the conditions of temperature and pressure. It is not possible to calculate or measure entropy in chemical reaction. Entropy change is highly a complex process. It is the reflection of disorderliness. For example, during respiration or zygotic development and during morphogenesis there is considerable change in entropy or disorderness. The entropy change has been used to guide the direction and equilibrium position of chemical reaction, namely, the change in free energy. The free energy is a form of energy capable of doing work under conditions of constant temperature and pressure. The relationship between the free-energy change of a reacting system and the entropy under conditions of constant temperature and pressure is given by

$$\Delta G = \Delta H - T\Delta S$$

where ΔG = free energy change of the system
ΔH = change in enthalpy
T = absolute temperature
ΔS = change in entropy

The free-energy change ΔG can be defined as that portion of the total energy change which is available to do work as the system moves towards equilibrium at constant temperature, pressure and volume.

The second law emphasizes that when one type of potential energy is converted into another, some of the potential energy is transformed into kinetic energy. Potential energy is not easily stored. It is converted to chemical energy in green plants which conserve less than 1 per cent of the light energy striking the earth.

3.3.2 The Cell and Thermodynamics

Normally cells are better equipped in handling heat energy efficiently than steam engine. All biological reactions take place in solution, with few exceptions at constant pressure and constant volume. Cells can function when they are at uniform temperature. Thermal gradient or temperature fluctuations may occur temporarily in cells when heat is produced. This thermal fluctuation is not essential for cells and no way connected to cellular function. Some chemical potential energy capable of doing work is continuously converted into heat during cellular reactions. This kinetic is used to maintain the temperature of the cell and it will not be converted into potential energy again. Consequently, free energy is required to supply to the cell from outside to sustain life.

QUESTIONS

1. Give an account of intra and inter molecular bonds in biological systems. (BU April 1999)
2. Give an account of forces stabilising molecular structure. (BU Nov. 1999)
3. Explain different types and nature of chemical bonds that stabilise the molecular structure.

SHORT QUESTIONS

1. Hydrogen Bond
2. Weak-Interaction
3. Ionic Bond
4. Covalent Bond
5. Van der Waals Force
6. Bonding nature in peptides
7. Bonding nature in nucleic acid

4

Analytical Techniques

4.1 THIN LAYER CHROMATOGRAPHY

Thin layer chromatography is very popular because of convenience, easy to perform and speed. This technique is very useful in quantitative analysis. The method is based on adsorption chromatographic principles.

Thin layer plates consist of a layer of adsorbent spread over an inert, flat support. This support is usually made up of glass or aluminium plates. A silica slurry, prepared in water (stationary phase) is spread on to a glass or aluminium plate as a uniform thin layer. A uniform coating can be accomplished by employing plate spreader, starts applying at one end of the plate and moving progressively to the other. The thickness of the slurry layer depends on the nature of separation, for example, slurry thickness of 0.25 mm is used for analytical separation and it may go even upto 5 mm. During slurry preparation, a binding agent such as calcium sulphate is added into the slurry to enhance the adhesion of the adsorbent to the plate. Plates are prepared by shaking a mixture of the adsorbent and the binding agent with an appropriate volume of water. The silica slurry coated plates are then allowed to dry to leave the coating of stationary phase. The drying is done by keeping the plates in an oven at 100-120° C. This entire process facilitates activation of adsorbent.

4.1.1 Sample Application

Samples can be applied to the plate as discrete spots or streaks with the help of capillary tubes or micropipette. The spot of sample is placed 2–2.5 cm from the edge of the plate. Sample spotting can be repeated to concentrate the material.

4.1.2 Development of Chromatogram

After spotting the sample, TLC plates are kept in the glass tank containing the development solvent to a depth of about 1.5 cm. Glass tanks are generally saturated with solvent vapour prior to separation. This equilibration process may be carried out for 1 hr. unless tank becomes saturated with solvent vapour. Irregular running of the solvent may be inevitable as it ascends the plate by capillary action, culminating in poor separation. Once equilibration is attained, the top cover plate is removed and thin layer plates placed vertically in the tank containing solvent system of 0.5–1 cm deep. In cover plated and completely covered by aluminium foil. The separation of the compounds begins as the solvent moves up the plate, kept under constant temperature. Separation in t.l.c. could be achieved between 20-45 minutes or maximum time may not reach 90 minutes.

4.1.3 Two Dimensional Separation

Separation of compounds in t.l.c. may not be satisfactory at many times. Therefore improvement in t.l.c. resolution of specific separations can be accomplished by adopting the technique of two-dimensional chromatography. In this process, the plate to be chromatographed is placed towards one corner of the plate as a single spot and the plate is developed in one direction and then removed from the tank and allowed to dry. The t.l.c. plate is then developed by another solvent in a direction right angles to the first development.

4.1.4 Detection

Detection of sample or component is carried out by spraying the plate with sulphuric acid (50%) in ethanol. This can be heated to see the appearance of compounds as brown spots. Certain UV light base compounds can be detected by ultraviolet light. The separated compounds show up as blue, green or black areas. Detection of amino acid requires ninhydrin reagent and the quantification of compounds in the spot can be assessed by one plate or off-plate quantification. On-plate quantification is done by the use of radiochromatogram scanning in the case of radio-labelled compounds. Densitometry is often popularly used to quantify the substance. Similarly, off-plate quantification may be carried out by scrapping of t.l.c. spot and eluting the compound using suitable solvent.

4.2 GAS LIQUID CHROMATOGRAPHY (GLC)

Gas liquid chromatography is a widely used method for the qualitative and quantitative analysis of a large number of compounds. GLC has gained popularity because of its high sensitivity, reproductibility and speed of resolution. Separation of low polarity compounds can be effectively carried out by g.l.c.

4.2.1 Principle

This technique is based on the distribution or partition of compounds between two physical states or phases *i.e.*, gaseous and liquid phases. A stationary phase — silicon grease is a stationary phase of liquid material which is supported on an inert granular solid.

The mobile phases like nitrogen and argon carrier gas are passed through the column at a flow rate of 40–80 cm^3 min^{-1}. The column temperature is maintained to provide a balance between peak retention time and resolution. The impelling force in GLC is the flow of nitrogen and argon gas through the column, supplied from gas cylinder, gas pressure in the region of 20 lb in^{-2} being common.

Once the insulated sample being carried away by the mobile phase gases, compound leaves the column they pass through a detector. This is linked in an amplifier to a recorder, which records a peak as a compound passes through the detector.

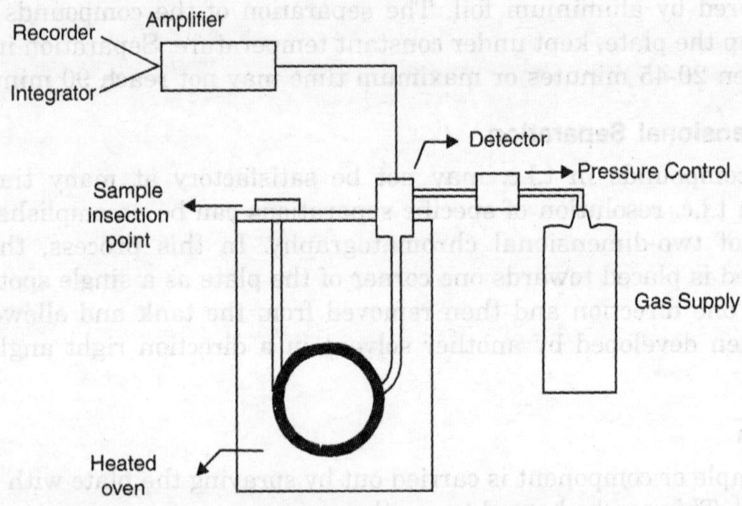

Fig. 4.1 GLC System

This stationary phase is packed into a narrow coiled glass or steel column of 1–3 meter long and 2–4 millimeter in diameter. The mobile phase consisting of inert gas is passed through this narrow column. Generally, certain inert gases like nitrogen or argon is used as mobile phase. The column is maintained at higher temperature by keeping it in oven. This is to volatilize the compound to be analysed.

Principally, compounds being separated is the difference in the partition coefficient of the volatilized compounds between the liquid and gas phases as they are carried through the column by the gas.

4.2.2 Sample Preparation and Application

Certain solvents such as ether, methanol or heptane are commonly used as dissolution of sample. The sample is injected on to the column using a micro-syringe. The sample quantity between 0.1 and 10 mm^3 of solution is injected. The test sample in the form of solution is introduced generally into the column at a temperature above the boiling point of the solvent. This strategy ensures rapid and complete volatilization of the sample.

4.2.3 Detector System

The most widely used detector in g.l.c. is the flame ionization detector. This detector is very sensitive and responds to almost all organic compounds. The sample concentration as low as nanogram can be detected by this detector. When the sample components leave the column, they are ionised in the flame associated with detector, resulting in an increased signal and passed to the recorder.

Another commonly used detector is the electron capture detector, particularly useful in detecting halogen-containing compounds. In addition, the thermal conductivity detector is the simplest detector, particularly sensitive to organophosphorus compounds. Therefore, many organophosphorus pesticide analysis is carried out by this detector.

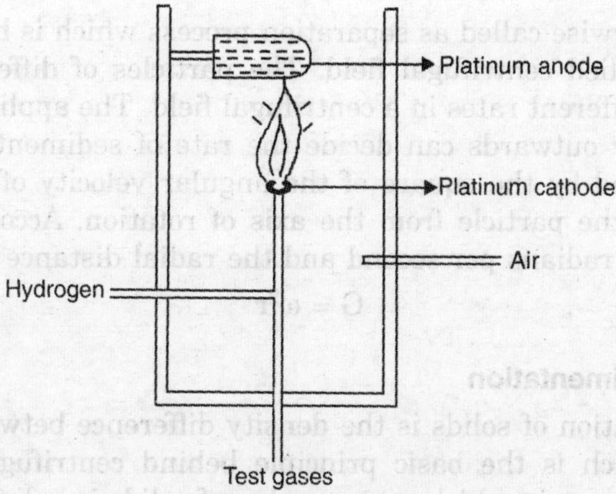

Fig. 4.2 A flame ionisation detector

Retention time parameter is considered to analyse the separated compound under standard conditions of temperature 1 gas flow. The time taken for a compound to emerge from a column is constant and is called the retention time. The column efficiency (E) is determined by following expression:

$$E = 16\left(\frac{y}{x}\right)^2$$

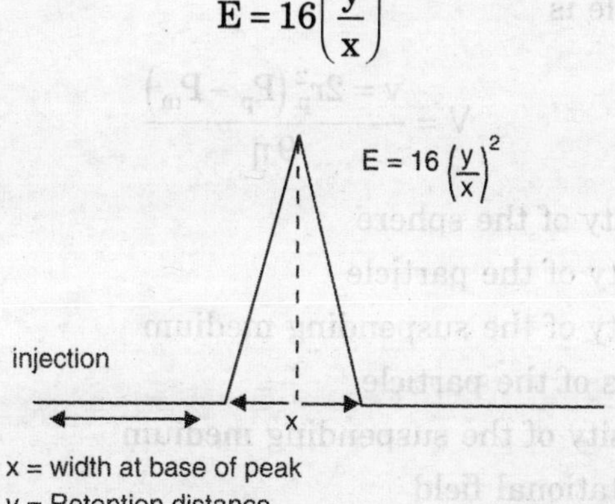

4.2.4 Applications of GLC

1. G.l.c. is widely used in the separation of biomolecules like amino acids, fatty acids, esters and steroids.
2. Several compounds like aromatics which are native to the plants as pharmaceutical importance can be separated.
3. General analysis of high boiling materials.
4. Separation of volatile amines.
5. Separation of many hydrocarbons.
6. Analysis of polar materials.

4.3 CENTRIFUGATION

Centrifugation is otherwise called as separation process which is based on the behaviour of particles in an applied centrifugal field. The particles of different density, size and shapes, sediment at different rates in a centrifugal field. The applied centrifugal field (G) being directed radially outwards can decide the rate of sedimentation. The centrifugal field can be determined by the square of the angular velocity of the rotor (ω) and the radial distance (r) of the particle from the axis of rotation. According to the equation, velocity of the rotor is radians per second and the radial distance in centimeter.

$$G = \omega^2 r$$

4.3.1 Centrifugal Sedimentation

The centrifugal separation of solids is the density difference between the solids and the surrounding fluid which is the basic principle behind centrifugal separation process. Sedimentation is a process in which a suspension of solids in a liquid undergoes gradual settling down under the influence of gravity. However, centrifugation process is going to trigger the process of sedimentation aided by centrifugal force. When a solid particle moves through a viscous medium, its velocity is affected by two opposing forces like gravitational or drag forces.

The rate of sedimentation of a spherical particle is dependent not only on applied centrifugal field, but also upon the density and radius of the particle and the viscosity of the suspending medium. The Stoke's law explains this relationship and the sedimentation of a rigid spherical particle is

$$V = \frac{v = 2r_p^2 (P_p - P_m)}{9\eta}$$

where
v = velocity of the sphere
p_p = density of the particle
p_m = density of the suspending medium
r_p = radius of the particle
η = viscosity of the suspending medium
g = gravitational field
$\frac{2}{9}$ = shape factor

4.3.2 Preparation Centrifugation

Preparation centrifugation is conceived with the isolation of biological material. One of the main approaches in preparation centrifugation is the differential centrifugation, which is based on the differences in sedimentation rate of particles of different size and density. Once centrifugal field is applied in step with manner, the material to be separated is resolved into a number of fractions in centrifuge tube. After centrifugation, pellet and supernatant are separated and pure pellet can be recovered by repeated washing. During differential centrifugation, particles of varying sizes and densities are distributed homogeneously at the commencement of centrifugation. Large sized particles settle first at the bottom followed by accumulation of medium sized particles. Pure solvent can accumulate at the top of the supernatant.

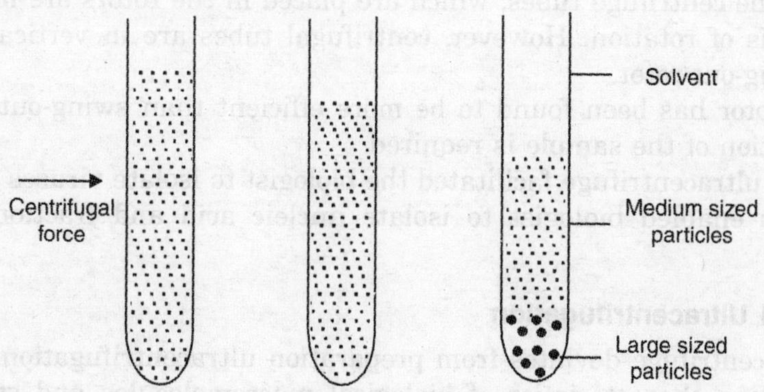

Fig. 4.3 Differential centrifugation

4.3.3 Density Gradient Centrifugation

In density gradient centrifugation, sample is layered on top of a continuous density gradient. Sedimentation of the particles is allowed to occur until the buoyant density of the particle and the density of the gradient are equal.

Nature of gradient material in density gradient centrifugation is crucial. One of the most convenient gradient material in density gradient is sucrose solution. Alternatively Ficoll may be used as substitution to the sucrose for the separation of molecules. Density gradient technique is categorised into two groups like discontinuous density gradient and the continuous density gradient technique.

In discontinuous density gradient technique, solutions of decreasing density are placed over each other in the centrifuge tube usually by pipette. The tube is then centrifuged under suitable conditions. However, continuous density gradient technique requires specialised instrument known as gradient former. This consists of two cylindrical chambers of identical diameter. Sample application to the gradient is carefully monitored. Separation of many biomolecules like proteins, enzymes, ribosomes and linear gradients are selected as suitable choice. However, concave gradients have been found to be satisfactory. Separation of cells or organelles require discontinuous gradients.

4.4 ULTRACENTRIFUGATION

Ultracentrifugation is categorised into preparation ultracentrifugation and analytical ultracentrifugation. Generally, high speed centrifuge is used with reachable speed upto 25,000 rev. min^{-1} and maximum centrifugal force of 89000 g. These instalments are used to collect microbes, large cellular organelles, cellular debris. Ultracentrifugation has maximum speed upto 75,000 rev. min^{-1} and even upto 1,00,000 rev. min^{-1}.

4.4.1 Preparation Ultracentrifugation

Preparation centrifugation can reach the maximum speed of 75,000 rev. min^{-1} and can produce centrifugal force upto 503.500 g. Entire rotor chamber is refrigerated to minimise generation of heat produced by friction. Rotors in ultracentrifugation are so called angle rotors because the centrifuge tubes, which are placed in the rotors are maintained at an angle to the axis of rotation. However, centrifugal tubes are in vertical position when placed in a swing-out-rotor.

The angle rotor has been found to be more efficient than swing-out-rotor when the total sedimentation of the sample is required.

Preparation ultracentrifuge facilitated the biologist tc isolate viruses in pure form. In addition, it has enabled biologist to isolate nucleic acid and fractionate subcellular organelles.

4.4.2 Analytical Ultracentrifugation

Analytical ultracentrifuge deviates from preparation ultracentrifugation by the study of only sedimentation characteristics of biological macromolecules and structures rather than the collection of particular fraction. In addition, presence of designed motors and detector system helps in constant monitoring the process of sedimentation of the material in a centrifugal field.

Analytical centrifuge can reach the speed upto 90,000 rev. min^{-1} and generates centrifugal fields upto 500,000 g. This centrifuge consists of a unique type of rotor called elliptical rotor coupled to a high speed drive unit. The rotor chamber is refrigerated and capable of holding two cells such as analytical cell and counter balance counter poise cell which is held vertically in the rotors at all times. For the observation of sedimenting material in the cell an optical system is placed in the centrifuge. This sedimentation process can be monitored by using ultraviolet absorptions or by difference in refractive index. When refraction index is used, the light passes through a transparent fluid with different density zones, in boundary between that zones refracts the light. In sedimenting materials in an analytical cell, a boundary between leaves of light particles will act like refraction lens, resulting in the formation of a peek in the photographic plate in detector system.

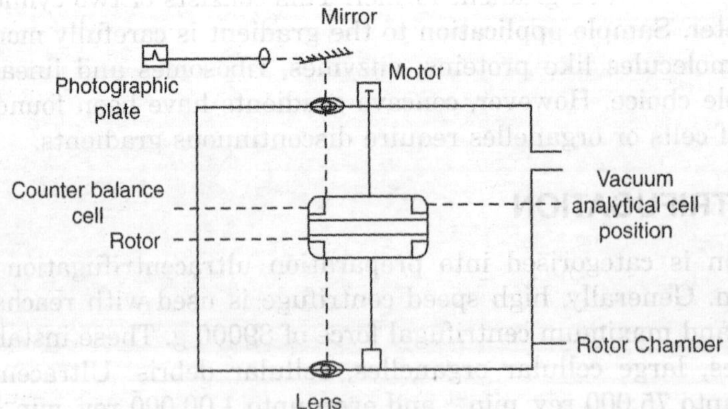

Fig. 4.4 Analytical ultracentrifuge

4.4.3 Applications of Ultracentrifuge

- Isolation of viruses, nucleic acids.
- Isolation and purification of subcellular organelles.
- Determination of molecular weight of biomolecules by analytical ultracentrifugation, ranging from two hundred to several million daltons.
- Due to the availability of large centrifugal fields to ultracentrifuges.
- Analytical ultracentrifuge is widely used into estimation of purity of macromolecules especially purity and the DNA preparing viruses and proteins.
- It is possible to detect the conformational changes in macromolecules, in particular double standard nature of DNA which can be either linear or circular in nature. It is even possible to detect conformation changes in allosteric enzymes.

5

Spectroscopy

Positive implications of spectroscopy in biological investigation is well established and has been used for several decades by investigators to elucidate the structure of compounds. One of the major efficiency of this technique in studying the structure of a compound is that it does not damage or degrade the molecule. In addition, even minute quantities of material can be precisely detected and assayed by this technique. Spectroscopy is indeed indispensable in the elucidation of structure of wide array of biological compounds.

5.1 BASIC PRINCIPLES

Light is an electromagnetic radiation consisting of a stream of packets of energy called photons or quanta. The amount of energy in each quantum determines the wavelength of the radiation.

Absorption of quanta of light energy can displace the position of electron in an atom. Normally in the ground state of the atom, electron is in the lowest energy level in accordance with the laws of thermodynamics. Raising of an electron from the energy level in the ground state to an excited state requires additional energy. Thus, elevation of electron energy is accomplished by the absorption of discrete quantum of energy. The electron is now in the excited state and it is equivalent to that involved in their transition. In the same line of principle the excited electron when falls back to its lower energy ground state, it emits radiation of specific wavelength (Fig. 5.1).

The electron occupies a status of new energy level when bonding takes place between two atoms to form a molecule.

In a molecule each ground and excited state is subdivided into number of energy sublevels as a status of molecular spectrum and are usually seen as band spectra. In a molecule, electrons change their energy levels when they absorb photons or quanta of radiation; it is referred as absorption spectra, or if electrons change their energy level emitted by the molecule; it is referred as emission spectra.

$$E = E_1 - E_2 = h\nu$$

where E = The energy of radiation absorbed or emitted by the molecule

E_1 = Electron energy in the ground state (original level)

E_2 = Electron energy in the excited state (Final level)
h = Planck's constant = 6.63×10^{-34} Js
ν = Frequency of the radiation in hertz = c/λ

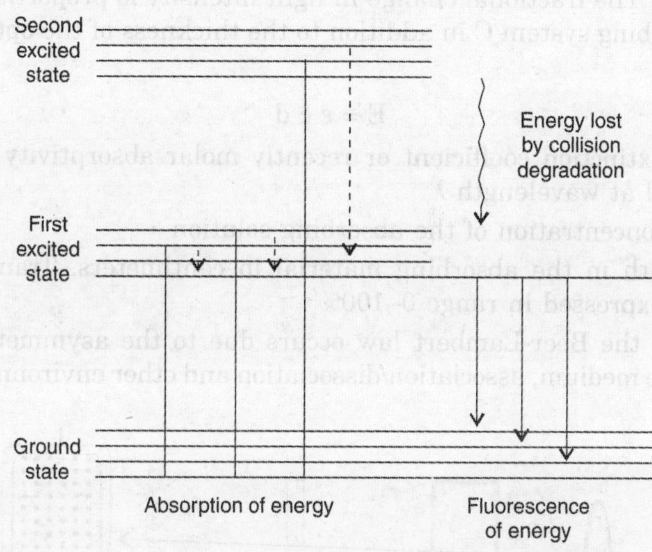

Fig. 5.1 Energy levels and transition of electrons

Hence, spectrum can be depicted as a scale of amount of energy absorbed or emitted by a system against wavelength. Generally, molecules absorb radiation of light. Molecular interaction with radiation having a number of ranges of wavelength is responsible for the generation of spectra in a number of distinct regions. These distinct regions can be studied by specialised instruments as they could yield valuable information.

5.2 BEER-LAMBERT LAWS

These two are the early empirical laws governing the absorption of light by molecule. Beer law relates the absorption to the concentration of absorbing solute. Lambert law relates the total absorption with reference to the path of light. Now they are conveniently used as the combined Beer-Lambert law. Lambert law states that the extinction is proportional to the concentration of the absorbing substance and to the thickness of the layer.

The proportion of the radiation passing through the absorbing medium is called the transmittance (T)

$$T = \frac{I}{I_0} \text{ where}$$

I_0 = Intensity of the incident radiation
I = Intensity of the transmitted radiation
The term absorbency is used instead of old optical density (OD).

In this equation I_0 is the intensity of light, I is the transmittance *i.e.*, log $(I/I_0) = A$. *A* is the absorbance. Since solutions are used in biology and chemistry, this equation was modified by Beer, so that it can be applied to the solutions. Beer noticed that when light passes through solution, in which the chemical form of the substance does not change with concentration. The fractional change in light intensity is proportional to the concentration of the absorbing system C in addition to the thickness of the optical path (sample) (Fig. 5.2).

$$E = \varepsilon c d$$

where ε = Molar extinction coefficient or recently molar absorptivity for the absorbing material at wavelength λ

c = Molar concentration of the absorbing solution

d = light path in the absorbing material in centimeters. Transmittance is commonly expressed in range 0–100%

Deviation from the Beer-Lambert law occurs due to the asymmetric distribution of the molecules in the medium, association/dissociation and other environmental parameters.

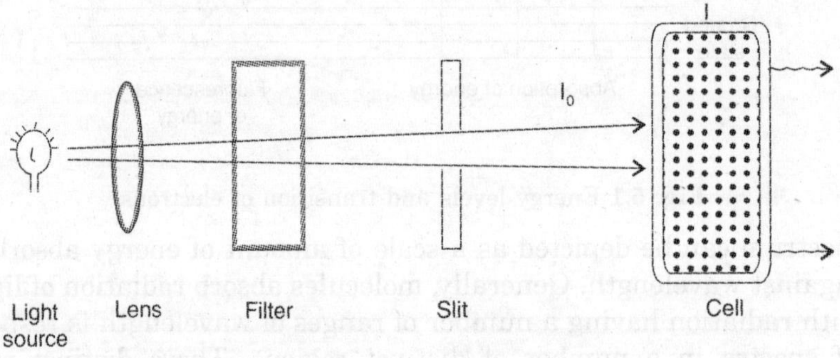

Fig. 5.2 Optical depiction of Beer-Lambert Law

5.3 COLORIMETRY

Colorimetry operates on the principles based on Beer—Lambert law. This simple operating system contains fixed wavelength filters, light source, lens, slit or aperture, photosensitive device, amplifier and meter (Fig. 5.3).

5.3.1 Principle

When a light of monochromatic or heterogenous falls upon a homogenous media, a portion of the incident light is reflected, a portion is absorbed within the medium and the remaining is transmitted. If the intensity of the incident light is expressed by I_0, that of the absorbed light by I_a, that of the transmitted light by I_t, and that of the reflected light by I_r then

$$I_0 = I_a + I_t + I_r$$

colorimetry is concerned with the determination of the concentration of a substance by measurement of relative absorption of light with respect to a known concentration of the substance. In colorimetry, white light is generally used as a light source, and determination is usually made with photoelectric cell. In colorimeters, light contained within a comparatively narrow range of wavelengths furnished by passing white light through filters of coloured glass, gelatin etc. is employed, transmitting only limited spectra regions. Colorimeters are also called filter photometers.

As per Lambert's law when monochromatic light passes through a transparent medium, intensity of the emitted light decreases exponentially as the thickness of the absorbing medium increases arithmatically. As per Beer's law the intensity of a beam of monochromatic light decreases exponentially as the concentration of the absorbing substance increases arithmatically. In the fundamental equation of colorimetry it is often spoken as the Beer-Lambert law.

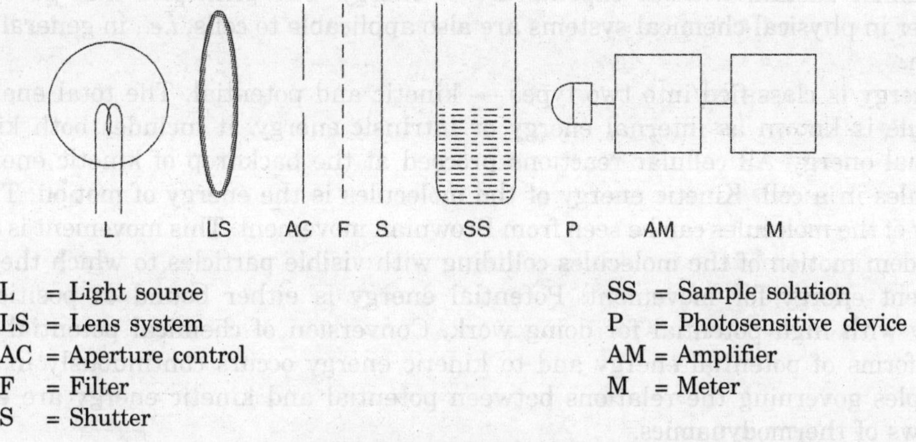

L = Light source
LS = Lens system
AC = Aperture control
F = Filter
S = Shutter
SS = Sample solution
P = Photosensitive device
AM = Amplifier
M = Meter

Fig. 5.3 Colorimetry

Many substances which do not exhibit sufficient extinction coefficient in the visible region will react with some other reagents to produce coloured product. This is the basis of assaying such substances. The chromophores or colour is produced under standard conditions from the known amount of substance used in the blank in the reference cuvette extinction of this sample is measured and the blank contains all the reagents except the sample being estimated. The extinction reading is set to zero for blank. The measured extinction is plotted against the concentration of the test sample. This is called as calibration curve. The colorimetry is widely used in various fields to assay biologically important molecules as it produces high extinction with relatively small amount of material.

5.3.2 Applications

- Colorimetry is widely employed in diversified fields such as analytical biochemistry for the estimation of concentration of biological samples.
- Employed in clinical, pharmaceutical, food, beverage and water analysis.
- It is also used in analytical and agricultural chemistry.

5.4 PRINCIPLES OF THERMODYNAMICS

Thermodynamics deals with energy relationship of any reaction. Thermodynamics is bound by certain laws which are universal. All physical, chemical and biological processes are bound by these laws. An understanding of the basic concepts of thermodynamics is indispensable for approaching problems in bioenergetics.

The process of thermodynamics is either reversible or irreversible. The reversible process can be seen in chemical reactions whereas biological systems are unique examples for irreversible process. A system may be closed or open. A *closed* system is the one in which only energy can pass through the system and the system is totally a barrier to the matter. On the contrary in an *open* system, both energy and matter can pass through the system. Examples for closed and open systems are chemical reactions and biological systems respectively.

Cells are living physical and chemical systems that absorb, transduce and effectively utilise energy in maintaining the sustainability of life. Energy is the capacity to do work. Life cannot sustain without expenditure of energy. The principles that govern energy transfer in physical-chemical systems are also applicable to cells, *i.e.*, in general biological system.

Energy is classified into two types — kinetic and potential. The total energy of the molecule is known as internal energy or intrinsic energy. It includes both kinetic and potential energy. All cellular reactions proceed at the backdrop of kinetic energy of the molecules in a cell. Kinetic energy of the molecules is the energy of motion. The kinetic energy of the molecules can be seen from Brownian movement. This movement is attributed to random motion of the molecules colliding with visible particles to which they transfer sufficient energy for movement. Potential energy is either bound or positional. It is energy with high potential for doing work. Conversion of chemical potential energy to other forms of potential energy and to kinetic energy occurs continuously in cells. The principles governing the relations between potential and kinetic energy are encased in the laws of thermodynamics.

5.4.1 Laws of Thermodynamics

5.4.1.1 First Law of Thermodynamics (Enthalpy)

The first law of thermodynamics is the principle of energy conservation. In any process (physical, chemical or biological) the total energy of the system and its surrounding remains constant. Energy is neither created nor destroyed. However, it may undergo transformation from one form to another, for example, chemical energy may be transformed into thermal, radiant or mechanical energy; this law is associated commonly as Enthalpy (H).

$$\Delta H = \Delta U + P\Delta V$$

where U = internal energy
PV = pressure, volume
H = Enthalpy
ΔH = change in enthalpy
ΔU = total energy change
ΔV = change in volume

Endothermic and exothermic reactions can be explained on the basis of enthalpy change. If enthalpy change exceeds zero ($\Delta H > 0$), then energy enters into the system.

This reaction is called endothermic whereas if enthalpy change falls below zero, then energy is lost from the system and the reaction process is called exothermic.

If $\Delta H > 0$ the process is endothermic (heat is absorbed)
$\Delta H < 0$ the process is exothermic (heat is released)

Enthalpy can be measured calorimetrically. In biological system, for instance, enthalpy change (ΔH) for the oxidation of glucose is –673,000 cal/mol of glucose at 20° C and 1 atmospheric pressure. As the temperature fluctuates, enthalpy changes are inevitable.

$$C_6H_{12}O_6 + 6O_2 \longrightarrow 6H_2O + 6CO_2$$

$$\Delta H_{293} = 673 \text{ kilo calories}$$

where enthalpy change ΔH is denoted by indication of appropriate absolute temperature in the kelvin scale (0° C = 273.1° K)
At 20° C = 273 + 23 = 293° K).

5.4.1.2 The Second Law of Thermodynamics (Entropy)

Entropy is used to denote measure of degree of disorder in any system is second law of thermodynamics. The entropy of the universe always increases until equilibrium is attained, at this point entropy is maximum under the conditions of temperature and pressure. It is not possible to calculate or measure entropy in chemical reaction. Entropy change is highly a complex process. It is the reflection of disorderliness. For example, during respiration or zygotic development and during morphogenesis there is considerable change in entropy or disorderness. The entropy change has been used to guide the direction and equilibrium position of chemical reaction, namely, the change in free energy. The free energy is a form of energy capable of doing work under conditions of constant temperature and pressure. The relationship between the free-energy change of a reacting system and the entropy under conditions of constant temperature and pressure is given by

$$\Delta G = \Delta H - T\Delta S$$

where ΔG = free energy change of the system
ΔH = change in enthalpy
T = absolute temperature
ΔS = change in entropy

The free-energy change ΔG can be defined as that portion of the total energy change which is available to do work as the system moves towards equilibrium at constant temperature, pressure and volume.

The second law emphasizes that when one type of potential energy is converted into another, some of the potential energy is transformed into kinetic energy. Potential energy is not easily stored. It is converted to chemical energy in green plants which conserve less than 1 per cent of the light energy striking the earth.

5.4.2 The Cell and Thermodynamics

Normally cells are better equipped in handling heat energy efficiently than steam engine. All biological reactions take place in solution, with few exceptions at constant pressure and constant volume. Cells can function when they are at uniform temperature. Thermal gradient or temperature fluctuations may occur temporarily in cells when heat is produced. This thermal fluctuation is not essential for cells and in no way connected to cellular function. Some chemical potential energy capable of doing work is continuously converted into heat during cellular reactions. This kinetic is used to maintain the temperature of the cell and it will not be converted into potential energy again. Consequently free energy is required to supply to the cell from outside to sustain life.

5.5 VISIBLE ULTRAVIOLET SPECTROSCOPY

When a compound absorbs certain amount of light against wavelength it exhibits characteristic absorption spectrum. The coloured compound will have one or more extinction (absorption) maximum in the visible region of the spectrum, *i.e.*, between 400–700 nm. One of the main reason for absorption spectra in visible and ultraviolet region (190–390 nm) is due to the energy transition of electron in the molecule. Normally π-bonding electron of carbon-carbon double bonds and lone pair of oxygen and nitrogen atoms are involved. A small part in a molecule independently gives rise to absorption spectrum. In a classical example, double bonds in carboxyl group lowers the energy indispensable for transition of electron and causes an increase in the wavelength at which chromophore absorbs. This phenomena is known as bathochromic effect. At this juncture, if wavelength is decreased it is referred as hypsochromic effect. Similarly hyperchromic effect is referred as increase in absorption (extinction) and hypochromic as decrease in absorption.

5.5.1 Principle

Spectrophotometry (visible and UV) works on the principles of Lambert and Beer's law in which when light of monochromatic or heterogenous impinges upon a homogenous medium, a portion of the incident light is reflected, a portion is absorbed within the medium, and the remaining light is transmitted. If the intensity of the incident light is expressed by I_0, that of the absorbed light by I_a, that of the transmitted light by I_t and that of the reflected light by I_r, then

$$I_0 = I_a + I_t + I_r$$

5.5.2 Description of Instrument

The optical system of simple spectrophotometer, i.e., visible and ultraviolet spectrophotometer consists of monochromators, cuvettes, photocells, slits and recorder. These are described as follows. (Fig. 5.4 and 5.5)

5.4.1.2 Light Source

The light source is usually a tungsten lamp for the visible spectrophotometer and either a hydrogen or deuterium lamp is used in ultraviolet spectrophotometer.

5.5.2.2 Monochromators

Monochromator is an optical system which can produce parallel beam of radiation of single wavelength (monochromatic radiation) from a multiwavelength source of radiation.

The monochromatic (single wavelength) radiation is possible due to refraction by a prism or diffraction by a grating. In visible spectro-photometer, prism is made up of glass. However, it is made up of quartz or silica in UV spectrophotometer since glass absorbs at wavelength below 400 nm. Diffraction grating monochromators comprises of a series of ruled lines on a transparent base. A series of overlapping spectra is formed by diffraction of white light. In a monochromator, prism can pre-select a portion of spectrum of light which is then diffracted to obtain monochromatic light.

5.5.2.3 Cuvettes

Cuvette is an optically transparent vial or cell in which the material under study is dissolved in a suitable solvent and placed to check absorbancy. The glass cuvette and quartz cuvette is used in visible and ultraviolet spectrophotometer respectively. Usually in a spectrophotometer the cuvette holder can hold between two to four cuvettes. The cuvette is used as reference cuvette and test cuvette. The test cuvette is used to set spectrophotometer to zero absorbancy. Absorption cuvette is always positioned in the light path in an instrument. The optical path length of cuvette is 1 cm and requires 2.5-3.0 cm^3 of sample for reading.

5.5.2.4 Photocells

Photocells transform light radiation into electrical energy which is then amplified, detected and recorded. When photons fall on a metal surface in vacuum condition, emission of electrons takes place in proportion to the intensity of radiation. The ejected electrons are then attracted by positive electrode causing flow of current, in turn generating potential difference across a resistor in the system. The absorption of light is accurately measured after electronically amplifying this potential and balancing it against a potentiometer calibrated directly in extinction. Unit sensitivity of photocells to photon radiation is as well recorded. Typical photocells are sensitive to wavelength of about 400 nm and are insensitive to wavelength above 550 nm.

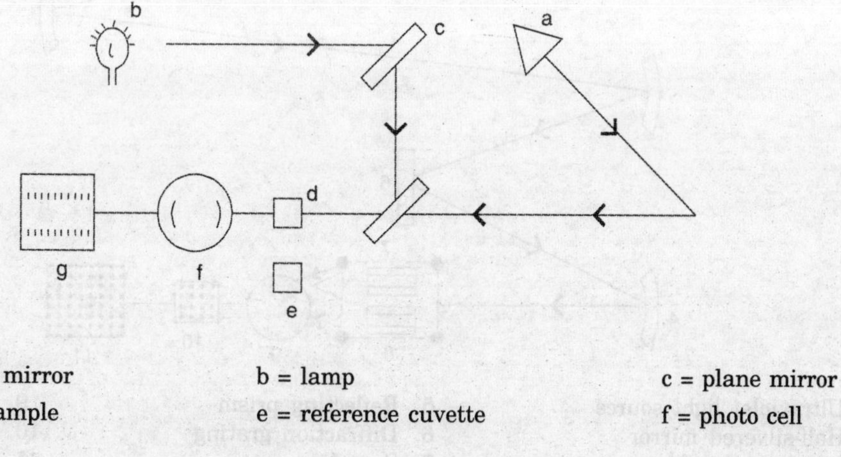

Fig. 5.4 Simple Spectrophotometer

5.5.2.5 Slits

The variation of slit width ensures adequate radiation reaching the photocell and the absorbancy depends upon the slit width. Zero absorbancy can be achieved in most spectrophotometers by adjusting the slit width. It is due to its effect on the band width and variation in the sensitivity of the photo cell with wavelength. The narrowest slit width is highly preferred to obtain reliable data in spectrophotometer.

5.5.2.6 (Read Out)

Recorder is able to scan the spectrum as well as measure variation in extinction at predetermined wavelength with time. The better recorder can scan the spectrum at varying speeds.

5.5.3 Overview of Operation

In spectrophotometer, light from the tungsten filament lamp (Deuterium in UV) is focussed on to the entrance slit by condensing lens. The light from the slit is collimated and

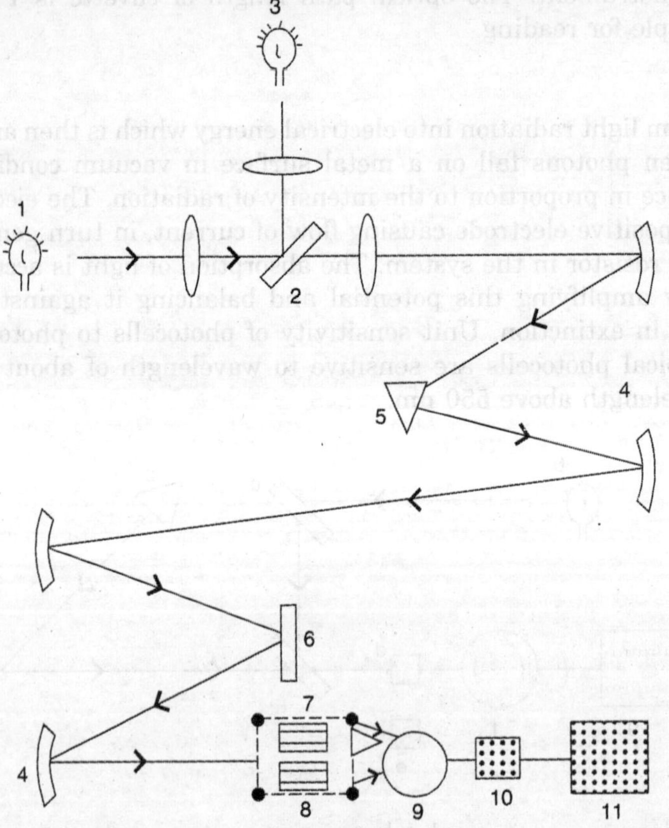

1. Ultraviolet light source
2. Half-silvered mirror
3. Visible light source
4. Mirror
5. Reflecting prism
6. Diffraction grating
7. Sample cuvette
8. Reference cuvette
9. Photo cell
10. Amplifier
11. Recorder

Fig. 5.5 UV Spectrophotometer

directed on to the grating which disperses the beam and the spectrum obtained as a result is focussed on to the exit slit. Various wavelengths are scanned by rotating the grating over its axis by rotating the wavelength disc which is coupled to the grating mounting. The monochromatic light is isolated by the exit slit passing through blank, standard or sample placed in cuvette and the transmitted light falling on a photo diode. The output signal from the photo diode is amplified, processed read to transmittance (T), Absorbance (A) and concentration (C) as selected.

5.6 FLUORESCENCE SPECTROSCOPY SPECTROFLUORIMETRY

5.6.1 Principle

This spectroscopy is used to record both excitation and emission spectra. When a molecule absorbs radiation it emits radiation of a longer wavelength. This phenomenon is known as Fluorescence. Thus a compound emits visible light after absorbing radiation in the ultraviolet region. Fluorescence is the emission of light that causes a molecular transition from the excited state to the ground state without alteration in electron-spin orientation. More precisely most organic molecules are in ground state at room temperature. When these molecules absorb photons of light, radiation elevates electron to an higher energy state, *i.e.*, S_1, S_2 etc. is less than 10^{-14} sec. When electron is displaced from its ground state due to absorption, its energy is lost very rapidly. As a result energy of the excited molecules descends rapidly to that of minimal vibrational energy in the lowest excited state (S, V_0). Although most of the molecules absorb in the ultraviolet and visible regions, only a few organic molecules emit fluorescence. This is because energy emitted from these molecules after attaining ground state in less than 10^{-8} sec, giving rise to fluorescent peak, *i.e.*, fluorescence is the immediate emission of light from a molecule after it has absorbed radiation in less than 10^{-8} sec. (Singlet-singlet transition). This is in contrast to phosphoresence which is the delayed release of the absorbed energy resulting from a triplet-singlet transition. In both the phenomena of fluorescence and phosphorescence, the light emitted will be of a longer wavelength than the radiation absorbed. The radiation emitted due to fluorescence is very specific and characteristic to molecules studied.

5.5.2 Instrumentation

Fluorescence spectroscopy consists of components like (a) Mercury lamp or Xenon lamp as a source of excitation energy, (b) pair of monochromator, (c) cuvette, (d) a detector like sensitive photo cell to measure the photo-luminescence and (e) amplifier (Fig. 5.6).

Glass cuvettes are generally employed as sample containers at wavelength above 3200Å. Similarly silica cuvettes are employed when studies are conducted below 32000Å. In addition, adequate temperature control is indispensable for precise fluorimetric estimation. This is because of decrease in intensity of fluorescence of the sample when temperature drops from 30° C to 25° C.

When radiation from light passes through monochromator, it disperses the light and is able to generate monochromatic radiation. This monochromatic light excites the sample when it passes through it. The excited sample then emits fluorescence or phosphorescence which passes into second monochromator. The monochromator selects the wavelength corresponding to the maximum fluorescence which then strikes the detector and recorded in the recorder.

Spectrofluorimetry is several thousand times more sensitive than absorption technique. It is very precise at very low concentrations. Spectrofluorimetry is able to detect the compound at 100 picograms when compared to 100 micrograms in absorption spectrophotometry. This high sensitivity of detection is due to wide range of amplification of the current produced in the photocell circuit. Presence of two monochromators enhance the utilisation of spectral selectivity.

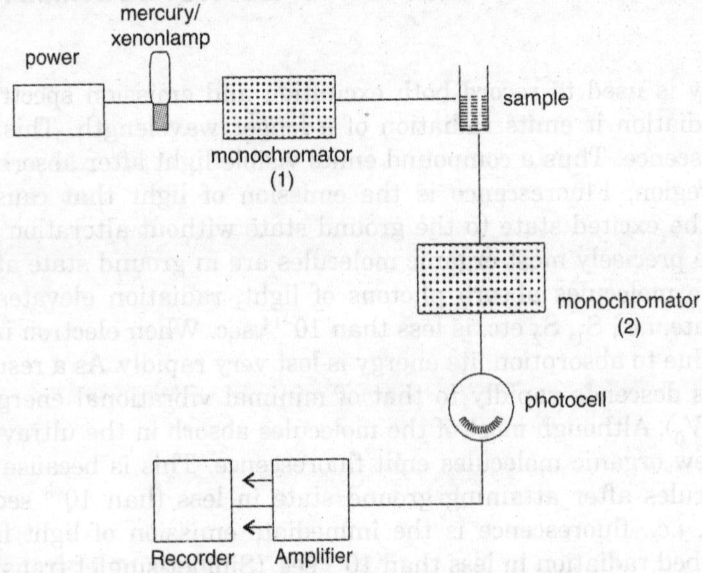

Fig. 5.6 Components of spectrofluorimeter

5.6.3 Applications

(a) Widely used in qualitative and quantitative analysis.
(b) Assay of vitamin B_1 in food stuff.
(c) Assay of NADH in mitochondria.
(d) Assay of this compound in microorganisms under various metabolic conditions.
(e) It is extensively used in the quantitative analysis of hormones like cortisol, oestradiol, and drugs like barbiturates in the blood.
(f) Widely employed in enzyme assay and kinetic analysis.
(g) Used in the studies on protein structure.
(h) Used in the assessment of NAD^+ and $NADP^+$ linked reactions.
(i) Spectrofluorimetry is also employed in the studies on membrane structure.

5.7 INFRARED SPECTROSCOPY (IR)

The infrared region of the electromagnetic region falls between visible and microwaves. It can be resolved into near infrared region (1 to 2.5μm), the infrared region (2.5 to 25μm) and the far infrared region (25μm to 400 μm). Infrared spectrophotometers monitor mid infrared region, *i.e.*, 2.5 to 25 μm as it would provide most useful data regarding detailed structure of compounds. The rotational and vibrational frequencies of molecules can be studied in the infrared region of the electromagnetic spectrum. Thus, infrared spectroscopy is very much useful in the study of molecular rotation and vibrations. In addition, several polar interactions like hydrogen bonding, conformation change and hydration can be studied.

5.7.1 Principle

The atoms in a molecule exist in the state of continuous vibration and different vibrations of a molecule exhibit different frequencies. Vibrations of the molecule result in a change in dipole moment. More precisely charge displacement takes place in the infrared region. The region between 8-15.5 μm in infrared radiation is being termed as finger print region and is used to describe the characteristic infrared pattern observed in various groups of atoms in a molecule upon interactions with infrared radiation absorb different parts of incident radiation depending upon their resonance frequencies. Vibration of this group is felt by other regions of the molecule. The interaction of many different vibrations of various parts of the molecule can produce harmonic vibrations. These harmonic vibrations absorb in other regions of the spectra. This possesses a frequency expressing the integral multiples of fundamental bands. The fundamental and harmonic multiplicity of vibrations produces a highly complex absorption spectrum characteristic to the functional groups and also overall configuration of the molecule.

5.7.2 Instrumentation

The overall basic construction of infrared spectrophotometer can be comparable to the ultraviolet and visible spectrophotometer. Infrared spectrophotometer for analytical studies are double beam instruments. Following are the key components present in the infrared spectrophotometer.

5.7.2.1 Light Source

For radiation source, tungsten filament lamp is used to supply the near-infrared radiation. In addition high pressure mercury lamp can be a radiation source in infrared spectrophotometer. The incandescent radiation source have high energy output, maximum at wavelength of two micrometers.

5.7.2.2 Monochromators

Since glass is opaque to infrared radiation, potassium bromide or sodium chloride prism monochromators were used. However, present day monochromators are grating monochromators due to their resolution efficiency and linear dispersion with wavelength. Since sodium chloride or cesium iodide prism are used in the mid-infrared, quartz prism is employed in the near infrared. The filters aid in selecting the required wavelength and avoid rest of the unwanted radiation.

5.7.2.3 Sample Container

In order to irradiate the samples in the infrared, special cells made from silver halides are widely used. The quartz and glass containers can be used as they absorb infrared radiation. Samples in aqueous solution are not useful as water exhibits intense absorption bands like any other solvents. Hence, fine paste of sample in liquid paraffin is prepared or discs of sample in potassium bromide is used. The samples must be well protected from radiation source to avoid rise in temperature by 5° C per hour.

5.7.2.4 Detectors

These are heat resistant with an infrared transparent window. In addition the expensive fragile goloy cell can be used. However, it is pressure sensitive system which comprises of a sealed container of gas. The infrared radiation causes raise in temperature which in turn expands the gas. Generally two types of detectors are used—(a) quantum detectors and (b) thermal detectors. The semiconductor receptor in quantum receptors can monitor the radiation impinges on it by generating electrical signals.

A servo system of recording is employed since infrared detectors respond non linearly to energy.

5.7.3 Applications

Following are some of the important applications of Infrared spectrophotometers in biological field:

(a) Study of biological macromolecules and membranes
(b) Helpful in infrared gas analysis in detecting and measuring differences in the concentration of CO_2, CO and acetylene in biological samples
(c) In photosynthesis, it is used to study carbon dioxide metabolism and also during respiration in plants

5.8 RAMAN SPECTROSCOPY

Raman spectroscopy and infrared spectrum are basically similar, but the basic principles are fundamentally different. Both these spectrum however provide information on the vibrational and rotational frequencies of the molecule. The Raman spectrum is based on the light scattering phenomenon when compared to absorption of radiation by molecule in infrared spectrum.

5.8.1 Principle

Raman Spectrum is based on the vibrational and rotational frequencies produced when light is scattered by the molecule. Scattering of electromagnetic radiation takes place when a monochromatic incident radiation is passed through the sample in visible region of the spectrum. In the light scattering process only the electric waves are involved.

Scattering of electromagnetic radiation by molecule takes place in two ways. One is elastic scattering, another is inelastic scattering. In the case of elastic scattering, wavelength of the scattered radiation is not altered. In this, elastic collision between photon and the molecule no exchange of energy takes place (no change). Whereas in inelastic scattering there is a change in the wavelength of the scattered radiation. In this, inelastic collisions of photons with molecules some energy transfer takes place from photons to the molecule and vice-versa. This inelastic scattering of electromagnetic radiation (light) by a molecule is known as Raman effect. In addition, there should be change in the shape of the electronic cloud of the molecule.

In Raman frequencies (Raman Scattering frequencies) the molecule may absorb a portion of energy from the incident photon ($h\nu_0$), as a result scattered light exhibits lesser frequency than that of the incident radiation. This phenomenon is termed as a Stokes-Raman Spectrum. Similarly in Raman scattering some of the energy from the molecules in the excited state is transferred to the colliding photons (scattered radiation). This photon acquires additional energy and shows increase in the frequency of the scattered radiation, as a result it gives rise to anti-stokes lines. The difference between the incident frequency and a stokes/anti-stokes line will be equal to the vibration frequency of the molecule.

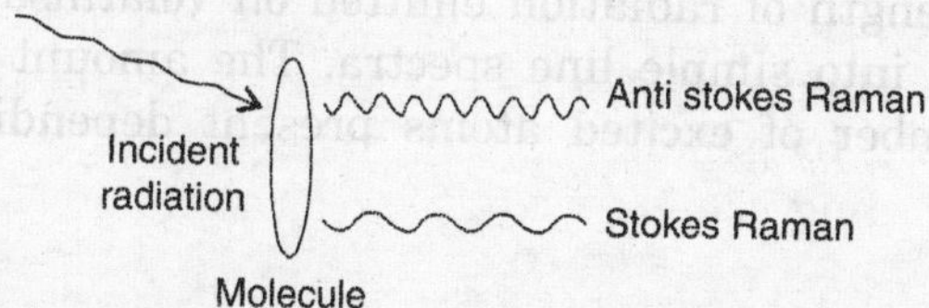

5.8.2 Instrumentation

The Raman spectrometer consists of (a) monochromatic radiation source, (b) sample container, (c) detector and (d) a recorder. Laser is the source for monochromatic radiation. The sample is placed in sample container fitted with an interference filter iris diaphragm, objective and polarisation analyser. The monochromatic radiation source laser beam is passed through interference filter, and iris diaphragm. The light is then well focussed by objective to spot on the sample. The aqueous nature of sample is taken in a quartz cell. The scattered light is then focussed by lens on to the entrance of a monochromator. The mirror collects the scattered light and reflects the rays scattered to the opposite side and place them to the collection optics. In addition, a double grating monochromator is used to produce higher resolution. The detector is connected to an amplifier.

In Raman spectrum the absorbance is measured and recorded as peaks. A spectrum is recorded graphically with the wave number or frequency of 200–350 cm^{-1}. Raman spectroscopy has far greater potential than infrared spectroscopy in explaining structural features of molecules. It is highly a preferred method as it is elusive and faster. Raman spectra can be obtained for compounds in any state like gaseous, liquid, solid and powdered form.

5.8.3 Applications

1. Used to study molecular vibration.
2. Useful in studying the functional groups and resonance.
3. It is also useful in elucidating the stereo-chemical features of molecule.
4. Raman scattering can be employed to probe polar and nonpolar environment.
5. Widely used to study structural and conformational features of biomolecules.
6. Raman Spectra can provide vital information on the peptide backbone, geometry of disulphide bridges.

5.9 ATOMIC ABSORPTION SPECTROSCOPY

5.9.1 Principle

Atomic absorption spectroscopy is based on volatilization of atoms in a flame causing them to absorb light of specific wavelength. Atomic absorption spectroscopy is abbreviated as AAS and in other words based on the absorption of light by free atoms. When the molecules come in contact with flame their atoms get dissociated due to vapourisation and absorb light of specific wavelength in their ground state. Absorption lines are characteristic to atoms present in the compound. Atomic absorption spectrophotometre measures the absorption of a beam of monochromatic light by atoms in a flame. The number of atoms present in light path can determine the absorption of energy.

In a AAS the wavelength of radiation emitted on volatilisation of element in flame may be readily resolved into simple line spectra. The amount of radiation absorbed is proportional to the number of excited atoms present depending on temperature and flame.

5.9.2 Instrumentation

The main components of an atomic absorption spectroscopy are (a) radiation source, (b) nebulisers and (c) detectors (Fig. 5.7).

5.9.2.1 Radiation Source

The most commonly used radiation source to produce a beam of radiation with a narrow band width are hollow cathode discharge lamp or white light with double monochromator. The hollow cathode lamp comprises of a tungsten anode and a cylindrical cathode in a sealed inert gas tube. Utility of discharge lamp is specific to the element.

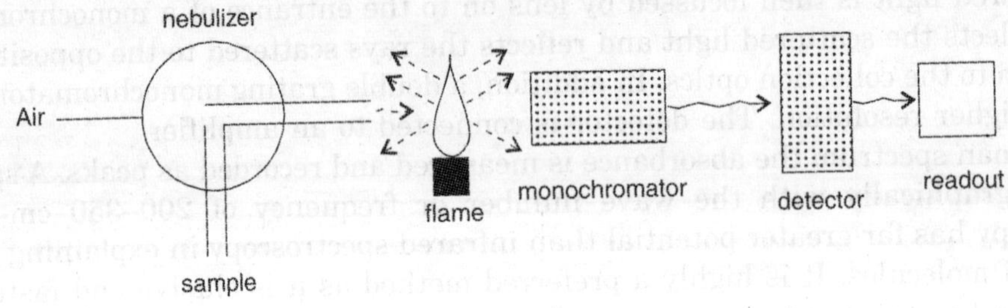

Fig. 5.7 Components of atomic absorption spectroscopy

5.9.2.2 Nebulisers

Nebulisers are otherwise known as atomisers. These are spray type in which stream of air passes over a capillary tube dipping into the test solution. A very tiny drop of test solutions are produced which are then passed with the help of direct injection system along with the air stream into the burner for volatilisation of atom.

Various gases are mixed with air to produce flame. These mixtures are air and natural gas, air and propane gas and air and acetylene gas, which can generate temperature from 1500° C to 2500° C.

5.9.2.3 Monochromators and Detectors: A simple filter, prism or grating monochromators are commonly used in AAS. Instruments with both single and double beam optics are used. Detectors are photocell. The read out for AAS in the visible region, *i.e.*, extinction values ranges from 0 – ∞. The desirable wavelength falls between 190–800 nm.

5.8.3 Applications of AAS

(a) Extensively used in biochemistry laboratory to study body fluid composition. The composition of blood, urine, saliva and cerebrospinal fluid can be assayed by AAS.
(b) Elements like sodium, potassium, calcium, magnesium levels are measured directly.
(c) In order to study copper, lead and mercury, these are extracted from biological fluids before being assayed.
(d) These are used extensively to study elements in plant after extraction from the plant sample.

5.10 X-RAY CRYSTALLOGRAPHY

Determination of molecular structure of matter is a challenging task. The physical method and other related techniques are based on various functional parameters like chemical bonds, bond angles, stereo chemistry and conformation of molecules. Besides, number of electrons and its configuration can throw light on the structural features of molecule.

Determination of three-dimensional structure of several biomolecules is aided by several powerful techniques. Elucidation of three dimensional features of proteins, DNA, enzymes and other biomolecules was possible due to X-ray diffraction, Nuclear Magnetic Resonance (NMR) and other related techniques based on their principles of light scattering, electric dipole movement, optical activity etc. These techniques are employed in highly diversified field. Effective utilisation of these physical methods in biological science have been implicated to the greatest extent in understanding our biomolecules. It has now turned out that lysozyme is the first enzyme whose three-dimensional structure has been determined and whose properties are understood at atomic level by X-ray diffraction technique.

X-ray crystallography is a novel and powerful technique in the elucidation of three-dimensional structure of matter including biomolecules. Structural features can be determined at molecular and atomic level. Therefore, X-ray crystallography has become invariably indispensable tool in biological science in the crackdown of biomolecular structure.

In 1912, Walter Friedrich and C.M. Paul discovered that if a beam of X-ray passes through a crystal it is turned in various directions. In this process some of the X-rays do not travel in straight line. After passing X-ray through a matter crystal, the transmitted X-ray falls on a photographic plate, where they produce dark central spot as well as pattern of fainter spot around it. The reason for this diffraction pattern is that X-rays are scattered or reflected by the electrons that form the outer part of each atom in the crystal.

In the electromagnetic radiation, X-rays occupy high energy region and exhibit wave characteristics. It is now extensively used to probe matter at atomic resolutions and also efficiently describes the arrangements of atoms and molecules periodically in crystals. In X-ray crystallographic studies the X-ray wavelength is used in the range of ≈1Å. In order to explain the structure of matter at high resolutions only, X-rays of short wavelength are required.

5.10.1 Principle

One of the main principle behind X-ray crystallography is the process of scattering of X-ray when it passes through a matter in a crystalline status. X-ray scattering is similar in principle to scattering in light microscopy as well as in electron microscopy in the back drop of fundamental differences.

5.10.2 Technique

The technique of X-ray crystallography consists of crystallising the molecule and then subjecting it to irradiation by a beam of X-rays and making a record of three-dimensional diffraction pattern.

In all X-ray diffraction procedures, the diffraction patterns of the scattered X-ray and transforming these diffraction patterns into an image (on photographic film) are involved (Fig. 5.8). For interpretation, "construction of image" is indispensable with the help of crystallographers and also from modern computers. This operational process is called "Fourier transformation".

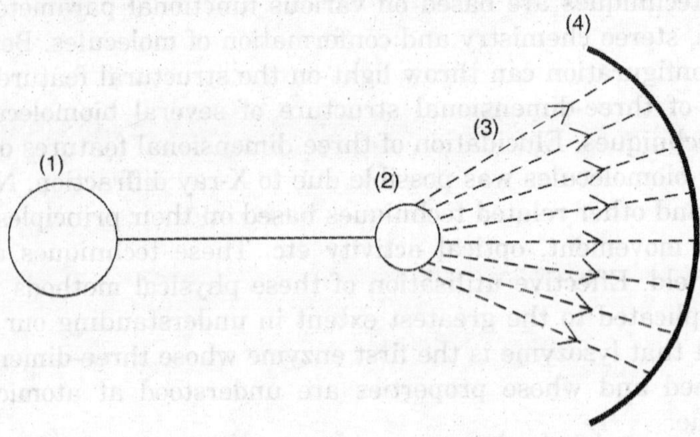

1. X-ray source
2. Crystalline sample (protein)
3. Diffracted beam
4. Photographic film

Fig. 5.8 Technique of X-ray crystallography

The term X-ray crystallography in practice, is the study of single crystals (matter/ biomolecule in crystalline states). The crystal acts as a three-dimensional diffraction grating, so that the waves scattered by them are confined to a number of discrete directions. In order to obtain a three-dimensional image of the structure, the intensity of the X-rays scattered in these different directions are measured. Since the X-rays are scattered by the electrons within the molecules, the image calculated from the diffraction pattern reveals the distribution of electrons within the crystal. The electron density is usually calculated at a regular array of points, and the image is made visible by drawing contour lines through points of equal electron density. Matter in single-crystalline state provides more information about its nature of atoms than in any other state. The matter in amorphous state provides only meagre information. On the contrary, powder diffraction methods are widely used to study the structure of certain other matter. The complete success in crystallography depends on the substance (sample to be studied) in the single crystalline state. Finally, the three-dimensional coordinates of each atom in the biomolecule (for *e.g.*, protein), as well as a number of which quantified its thermal vibration is obtained for further geometrical analysis.

5.10.2.1 Crystal lattice (unit cell)

A single crystal consists of a three-dimensional arrangement of atoms/molecules or clusters of molecules and it represented by a basic unit of structure called unit cell, which is repeated regularly in three-dimensional fashion called crystal lattice. There are several types of unit cell and combinations of symmetry operation (Fig. 5.9).

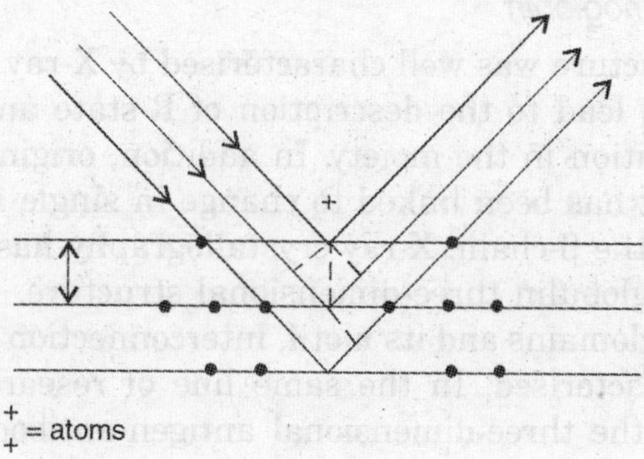

Fig. 5.9 Diffraction of X-rays

There are two limiting factors in the process of determination of biomolecule/matter structure through X-ray crystallography. One is the growth of the crystals which is poorly understood. Another is phase problem. One of the main obstacle in crystallography is the phase problem. It is the missing information in all X-ray diffraction data. Eliminating phase problem is central issue in the structure determination by X-ray diffraction. The mathematical method for transformation of diffraction pattern into an image of the object is called fourier transform. There are three methods of phase determination such as molecular replacement, Isomorphous replacement and finally anomalous dispersion. The molecular replacement is generally used crystallography to resolve the phase problem. In this strategic method, it is possible to locate the best orientation as position of a molecule in the unit cell whose structure has been determined. In the Patterson method or isomorphous replacement used to explain the structure and obtain the positions of the atoms in the unit cell. Anomalous dispersion is also used to resolve the phase problem in macromolecular crystallography.

5.11 CRYSTALLOGRAPHIC ELUCIDATION OF MOLECULES

5.11.1 Small Molecules

X-ray diffraction structural determination of small molecules provides resolution sufficient enough to resolve individual atomic position. Thus, it provides information of individual atoms in the unit cell coupled with the thermal vibration of the atoms. Several molecular parameters like bound length, its angles, stereochemistry, and conformational details of the molecules, besides hydrogen bonding, Van der Waals forces are calculated.

5.11.2 Macromolecules

Macromolecules such as globular proteins, enzymes and hormones can be crystallised into single-crystalline form due to persistent effort. Their crystallographic details can be elucidated similar to smaller molecules. However, the limit of resolution of globular macromolecules is 2–3Å. Due to this constraint, individual atoms or molecular shapes cannot be resolved. Thus, structure of macromolecules at low resolution (6–5Å) is done in the beginning to overcome phase problem and procure details on the size and shape of the molecule.

5.11.2.1 Structure of Haemoglobin

This macromolecular structure was well characterised by X-ray diffraction. Elucidation of structure of haemoglobin lead to the description of R-state and T-state of the molecule with changes in Fe^{2+} position in the moiety. In addition, origin of sickle cell anemia has been characterised and it has been linked to change in single amino acid residue due to single point mutation in the β-chain. X-ray crystallography has been widely employed in the discovery of immunoglobulin three-dimensional structure.

The immunoglobulin domains and its motif, interconnection of β-strands by disulphide linkages have been characterised. In the same line of research, X-ray crystallography provides information on the three-dimensional antigen-antibody complex which in turn could throw light on the nature of antigen-antibody reaction at molecular level and folding of polypeptide chain around crucial amino acid(s).

5.11.2.2 X-Ray Diffraction of Lysozyme

Lysozyme is the first enzyme whose three dimensional structure has been elucidated and its structural features are understood at atomic level. Its chemical make up is protein consisting of a single polypeptide chain of 129 amino acids. The 129 residue lysozyme molecule is cross-linked by disulphide bridges. The main polypeptide chain appears as a continuous ribbon of electron density running through the image. The amino acid residues in lysozyme have been designated by the number. The residues numbered 5 through 15, 24 through 34 and 88 through 96 forms three lengths of alpha helix. The conformation that was proposed by Pauling was found to be the most common arrangement of the chain in myoglobin. Even though the level of resolution achieved in the image was not enough to resolve individual atoms, many of the side chain characteristics of the amino acid residue were readily identified from their general shape.

The four disulphide bridges are marked by short rods of high electron density corresponding to the two relatively dense sulphur atoms within them. The six tryptophan residues were also recognised by the extended electron density produced by the large double-ring structure in their side chains. In addition, activity of lysozyme in attacking bacteria and breaking the linkages in bacterial cell wall by this enzyme was unravelled.

5.11.2.3 DNA helix structure

The proposed double helix model of DNA by Watson and Crick is actually based on X-ray diffraction photographs of B-DNA fibres. The proposed model refined several experimental data on the structure of DNA. It laid a basis for the explanation on replication and transcription of genetic material. The X-ray diffraction studies evidences the right-handed helical nature of B-DNA.

Applications

– X-ray crystallography methods have been immensely applied to elucidate the structure of small molecules such as nucleic acid component, amino acids, steroids, alkaloids, flavonoids and saccharides.

- It can be used widely for the structural determination of various macromolecules like proteins, enzymes, immunoglobulins, collagen, ATPases and types of nucleic acids — DNA and RNA.

5.12 NUCLEAR MAGNETIC RESONANCE (NMR)

Nuclear Magnetic Resonance (NMR) is a powerful technique for the investigation of chemical and physical properties at the molecular level. NMR is used to study molecular structure when the sample is placed in a strong magnetic field. It is then irradiated by a radio-frequency of electromagnetic wave. It has been used extensively as a novel tool for biological studies as well as for physical and chemical investigation.

5.12.1 Principle

Atomic particles (electrons, protons and neutrons) are spinning on their axes and these could generate magnetic fields having specific magnetic moments.

The nucleus in the atom consists of neutrons and protons. While the neutron does not have any charge, proton has one unit of positive charge. Their mass is about 1800 times larger than the electrons mass. The nucleons (both protons and neutrons) orbit around the centre of the nucleus and have orbital angular momentum. Both protons and neutrons have the same spin. In addition, there are energy levels for the nucleons (protons and neutrons) similar to the electronic levels (shells) in atoms. The protons will fill energy states by pairing themselves with spin-up and spin-down *i.e.*, the two protons are spinning in opposite direction. This pairing of protons produces cancellation of their spin angular momentum. The nucleus, however, exhibits angular momentum. The angular momentum of the nucleus is determined by the spin of the unpaired neutrons and protons and by the orbital angular momentum of the neutrons and protons. The combination of all these functions produces a very simple number for the value of the nuclear spin to this is designated by the letter I. The letter I is called the nuclear spin and the maximum detectable nuclear angular momentum is Ih.

When magnetic field is applied, the spinning nuclei tend to orient themselves to the field in the ground state. To this spinning nuclei, if additional energy is supplied in the form of electromagnetic radiation of a suitable frequency at right angles to the applied field, the nuclei absorbs energy and elevated to a higher energy state by reversing the spin against the field. As the energised nuclei cannot remain in the excited state for a long time, they tend to lose energy and falls back to the ground state. The energy absorption process starts again by the nuclei and falls back to the process, starts again by the nuclei and falls back to the ground state as this process is repeated several times. At this juncture, the nuclei are said to be resonating with radiation (NMR) and this frequency, where resonance takes place is called resonance frequency (Fig. 5.10).

The angular momentum is a physical quantity that describes the rotational motion of a body *i.e.*, a spinning nucleus has angular momentum. The angular momentum of the nucleus is important to NMR, Angular momentum is necessary for the nucleus to show its action when placed in the magnetic field. Since protons in the nucleus exhibit both spin and charge (positive) they behave like small magnets. The protons have magnetic moments like electrons, which several hundred times smaller than electrons magnetic moment. In the presence of magnetic field, proton and other nuclei with a nuclear spin valve of 1/2 can exist either in a low energy state with their nuclear spin aligned parallel with the field or in a higher energy state. The nuclear spin in a higher energy state is anti parallel to the field. The nucleus has to absorb appropriate quantum of energy in the electromagnetic spectrum to transform low energy state to high energy state.

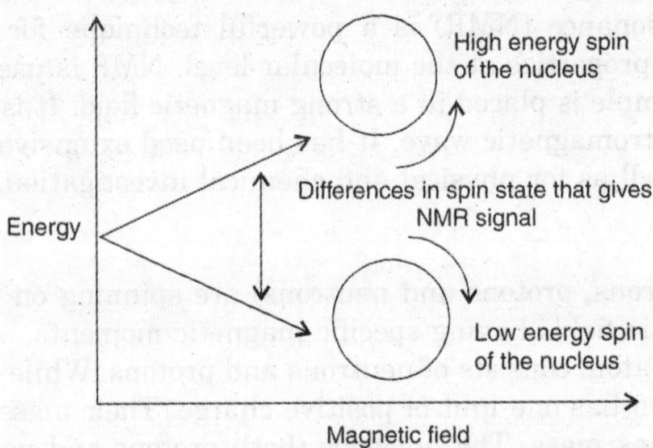

Fig. 5.10 General principles of NMR

Such nuclei when placed under magnetic field of several thousand Gauss are able to absorb radiation in the radio wave region of the electromagnetic spectrum. Energy from this wave is absorbed by each of the nuclei of the atoms in the molecule giving rise to the phenomenon known as nuclear magnetic resonance (NMR) according to their individual resonance frequency. The amount of energy which would be absorbed can be accurately predicted for every type of nucleus and any difference from this value is a measure of the effect of other nuclei present in the molecule. For instance, a proton or hydrogen (hydrogen nucleus is proton) nucleus attached to the carbon atom, would absorb energy from radiowaves at a slightly different frequency from one attached to a nitrogen atom. The most common element which can lend their nucleus to NMR are hydrogen, carbon, phosphorus and nitrogen. The studies which are commonly conducted in NMR are isotope of hydrogen (^{1}H). In addition, ^{13}C, ^{15}N, ^{31}P isotopes are also employed in biochemical studies as they possess a nuclear spin value of 1/2 the nuclei of other biological molecule. (^{12}C, ^{16}O, ^{32}S) possess zero spin and lack a magnetic moment and do not generate NMR signal. The nuclei of an element (proton) in different chemical environments give rise to distinct spectral lines. This is called chemical shift (Fig. 5.11). The nuclei in different chemical environments resonate at different frequency and the differences in resonance frequency are expressed as chemical shift.

The phenomenon of chemical shift arises due to the shielding of nuclei (proton) from the external magnetic field by electrons. This chemical shift is the basis of NMR spectroscopy and is an important parameter in structural studies by NMR. The chemical shift provides qualitative information and the intensity of any peak provides quantitative information about the groups present in the sample.

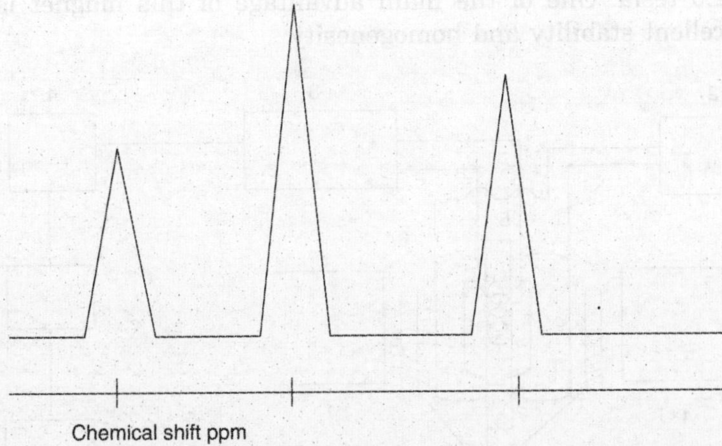

Fig. 5.11 Typical NMR spectral resonance as absorption bands/peaks

After the discovery of strong superconducting magnets several modified NMR versions have been using to elucide molecular structure. Some of these are fourier transform spectroscopy technique and the multi-dimensional NMR technique. These advances allowed analysis of molecules such as proteins which contains several hundred hydrogen atoms, which require interpretation of spectra containing hundreds of peaks in a proton NMR spectrum. Two-dimensional NMR techniques such as cosy is able to assign hundreds of peaks at a time. In recent years multi-dimensional nuclear magnetic resonance is employed as a structure determining technique. Since NMR probes the nuclei in the molecule it is more easily achieved when the molecule is in solution.

5.12.2 Instrumentation

The basic components of NMR spectrometer are:

(a) Electromagnets
(b) Radiofrequency
(c) Sample probe
(d) Sweep generator
(e) Detector and recorder

Electromagnets weighing 10^3 kg are used in the NMR studies. Permanent magnets can be used for NMR imaging if they are big enough. Permanent magnets with field strength of about 0.5 to 0.39 (500 to 3000 gauss) have been designed for use in NMR imaging. For NMR studies super conducting magnets are used as an alternative to other resistive coil magnet. Super conducting magnet has an advantage of the zero resistance, certain materials have at very low temperature. At zero resistance a current will flow continuously without further power input. At this stage entire magnetic system is maintained at about 4° K. Most commercial NMR imaging magnets are permanent or super conducting magnets.

Permanent magnets are available with a field strength of about 0.05 to one, 0.3 tesla as they do not require cooling. Super conducting magnets are available with field strength of about 0.3 to 2.0 tesla. One of the main advantage of this magnet is its high field strength with excellent stability and homogenesity.

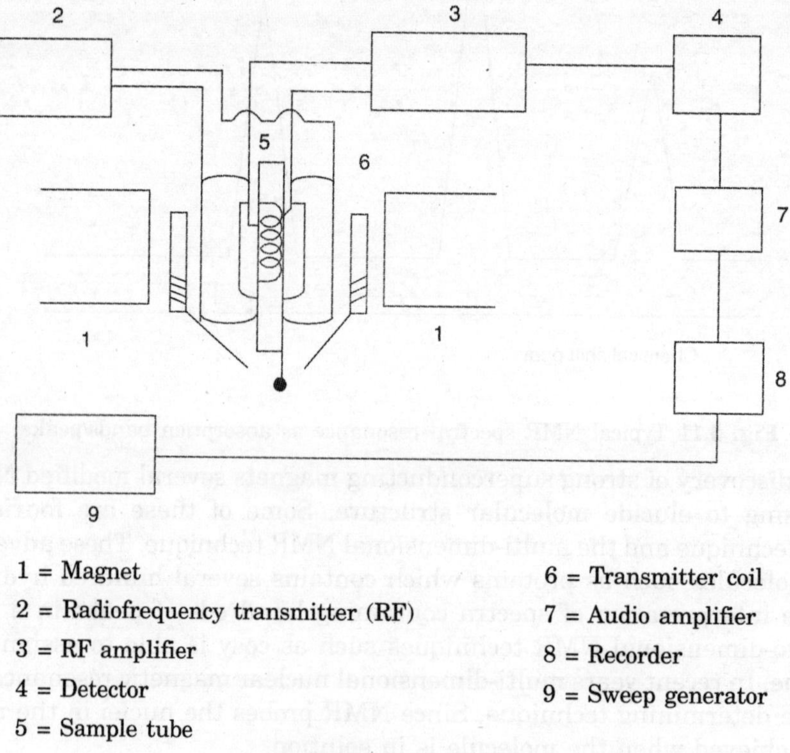

1 = Magnet
2 = Radiofrequency transmitter (RF)
3 = RF amplifier
4 = Detector
5 = Sample tube
6 = Transmitter coil
7 = Audio amplifier
8 = Recorder
9 = Sweep generator

Fig. 5.12 Apparatus for NMR

A radio frequency transmitter generates the monochromatic radiation. This is used to irradiate the sample. The radio frequency transmitter and a coil surrounding the sample vial is imposed at right angles to the magnetic field and has a frequency of 60 mHz. There are several types of radio frequency (RF) coils are used. One of the type of RF are saddle-shaped coil, another type is solenoidal coil. These are used to produce the magnetic field. Sample probe consist of a thin walled cylinder glass tube. The rotation of this tube can reach several hundred revolutions per minute. Besides single coil wrapped around the probe acts as a detector. The sample is taken in a glass tube which is placed between the poles of the magnet. A radio frequency source of 60 mHz is applied on the sample. The sample should be dissolved to a very high concentration in a solvent which lacks protons such as D_2O. When radio frequency is applied it causes small oscillating e.m. field on the sample of the energy difference between nuclear spin energy levels matches with the radiofrequency. This E.M. field induces a transition between energy levels when a transition occurs in sample. The oscillation in the field induces a voltage oscillation in the radio frequency receiver coil which is amplified and detected (Fig. 5.12).

5.12.3 Applications of NMR

1. Novel analytical method for chemical identification and characterisation of known and unknown structure of the molecules like proteins and nucleic acids.

2. It is used to observe molecular changes that occur in a molecule when its surrounding environment is changed. Protein conformation change can be observed in the reaction mechanism.
3. NMR is used to know changes occurring in the phosphate of nucleic acid under different chemical environment.
4. NMR has been implicated well to establish the form of D-glucose in solution.
5. Extensively used to study protein folding and dynamics.
6. The biomedical NMR deals with the application in cellular metabolism at the molecular level, biochemical energetics of living organism *e.g.*, changes in tissue concentrations of ATP and phosphocreatine in the forearm muscle before and after exercise.
7. Study differences in metabolism of cancer tissue from that of normal tissue.

ESSAY QUESTIONS

1. With the help of a neat labelled diagram describe a spectrophotometer. (BU May 1996)
   (BU April 1998)
2. With the help of a neat labelled diagram, describe fluorescence spectroscopy.
3. Explain the principles of spectrophotometer and give any two applications. (BU Oct. 1998)
4. Describe the principle, apparatus system of Atomic Absorption Spectroscopy.
5. What is Infrared? Describe the principle, operational mode of infrared spectroscopy.
6. Explain how molecular structure is determined using X-ray crystallography.
   (BU Oct. 1998)
7. Explain how X-ray diffraction technique is used in the determination of molecular structure.
   (BU Nov. 1999)
8. Describe the role of NMR in molecular structure determination.

SHORT QUESTIONS

1. Beer-Lambert Law (BU April 2000)
2. Colorimetry (BU April 2001)
3. Visible Spectrophotometer
4. UV Spectrophotometer
5. Monochromators
6. Fluoresence (BU April 2001)
7. Infrared Radiations (BU May 1996)
8. NMR (BU Oct. 1998)
9. NMR Imaging (Apri1 2000)
10. NMR methods in Spectroscopy (BU April 2001)
11. X-ray diffraction

6

Radioisotope Techniques

The principle aim of modern radiation biology is to study the effect of radiation on living systems and its applications to the benefit of mankind. Radiation biology has played a significant role in the progress of the biological field. Radioisotopes have been employed in the investigation of various cellular metabolic pathways such as protein synthesis, DNA replication, enzyme action etc. Besides, this technique is also being employed to study effect of radiation on nucleic acid and its subsequent effect on the nature cellular functions. In medical field, radioisotopic technique is widely employed in tumor therapy, immunotechnology and also as diagnostic tool in clinical biochemistry. Radiation techniques are mainly concerned with radioactive material, radioactive decay, screening methods and quantification of radioisotopes.

An atom is made up of positively charged nucleus, surrounded by cloud of negatively charged electrons. The protons and neutrons are the chief constituents of atomic nuclei. Protons are positively charged particles with their mass greater than that of electron. The number of orbital electron is equal to the number of protons present in the nucleus of atom. The total number of protons and neutrons determines the atomic weight or mass number of the element. Neutrons do not exhibit any charge and are neutral. The sum of protons and neutrons in the nucleus is called mass number. Some of the atoms of the element contain different number of neutrons hence a different mass number and are called *isotopes*. They are multiform of an element. They occur in nature. Nuclide is another word for isotopes. For example, carbon contains four isotopes — ^{11}C, ^{12}C, ^{13}C and ^{14}C. They exhibit different atomic weights due to differences in neutrons. ^{11}C has five neutrons, ^{14}C has eight neutrons. Similarly hydrogen (H) has three isotopes. They are ^{1}H, ^{2}H, and ^{3}H.

6.1 RADIATION

Light is a electromagnetic radiation. Although they have similar features, but differ in their energies *i.e.*, radio waves to visible light contains low energy photons, whereas ultraviolet, X-rays and γ-rays contain high energy photons. The energy level of photon is increased from ultraviolet to γ-rays when high energy photons interact with matter may lead to either excitation or ionisation. Ultraviolet containing high energy photons can induce excitation of the atom. However, it cannot trigger ionisation of the atom due to its excitation energy (less than 500 eV). On the contrary, ionising-radiations such as X-rays, γ-rays β and α-rays contain excitation energy (more than 500 eV) sufficient to cause ionisation (removal of electron from atom). Effect of ionisation is significant in biological system as they can cause alteration in the molecular structure, mutation in nucleic acids and even cell death. Molecular excitation by ultraviolet radiation may lead to carcinogenesis. Ionising radiations can be either pure radiation *i.e.*, x-ray and γ-rays or simply a particulate (α-rays, protons and neutrons).

The ionisation process takes place either directly or indirectly. Ionizing radiations such as α and β rays are ionising radiaticn directly while X-rays and γ-rays are causing ionising radiations indirectly.

6.1.1 Sources of Ionising Radiation

As far as source of ionising radiation is considered, both natural and man made are the main source. Some of the radiation like X-ray can be produced artificially. Natural radiation sources are cosmic rays and α-particles emitted by radioactive decay.

6.1.2 Types of Isotopes

The ratio of neutrons to protons in nucleus will decide on the stability of atom in nature. Isotopes are classified into two types — one is stable isotopes and another is unstable isotopes (radioisotopes). The unstable radioisotopes can be converted into another element with the concomitted emission of radioisotopes. Some of the commonly used radioisotopes are ^{3}H, ^{11}C, ^{14}C, ^{14}Na, ^{32}P, ^{35}S and ^{59}Co of which certain radioisotopes such as ^{14}C, ^{32}P and ^{14}C are extensively used in biological system.

6.2 RADIOACTIVE DECAY

Radioactivity is the emission of radiation by radioisotopes. Radioisotopes emit three different types of radiations. They are alpha (α) rays, beta (β) rays and gamma (γ) rays. Following are emissions of some of the radioisotopes.

6.2.1 α-rays

Radioisotope of uranium element ^{235}U emits α-rays and is converted into thorium. Generally isotopes of elements having high atomic numbers emit alpha particles (α). Emission of α-particles result in a significant lightening of the nucleus due to reduction in the atomic number and decrease in mass number. The utility of α-rays in biological system is insignificant and not frequently used. For example ^{226}Radium (^{226}Ra) decay by α-emission to ^{222}Radon (^{222}Rn). As the complex decay series begins it ultimately results in the formation of $^{206}_{82}pb$.

$$^{226}_{88}Ra \longrightarrow ^{222}_{86}Rn + ^{4}_{2}He^{2-}$$

$$^{235}U \longrightarrow ^{231}Th + \alpha\text{ - rays} \uparrow$$

6.2.2 β-rays

Radioisotope of carbon ^{14}C emits β-rays and undergoes conversion as below:

$$^{14}C \longrightarrow {}^{14}N + \beta\text{-rays}\uparrow$$

the emitted β-rays is negatron emission.

In this type of radioactive decay a neutron undergoes conversion to a proton by the ejection of a negatively charged beta (β) particle called a *negatron* (β-ve).

Neutron → proton + negatron

Negatron is high energy electron. As a result of emission the nucleus loses a neutron (electron) but gains a proton. Some isotope decay by committing emission of positively charged β-particles is referred as positron (β+ve). When proton is converted into neutron, positrons are emitted.

Proton → neutron + positron

positrons are unstable. As a result of positron emission, the nucleus loses a proton and gains a neutron.

6.2.3 X-rays and γ-rays

These are high energy photons. Gamma (γ) emission involves electromagnetic radiation. They show shorter wavelength than X-rays. In the production mode, these γ-rays are produced in nuclear reactions while X-rays are generated mainly because of bombardment of any heavy-element by high-speed electrons after atomic transition. Emission of γ-rays do not result in any change of atomic number or mass.

6.2.4 Rate of Radioactive Decay

It is a spontaneous process. Depending on the source, radio active decay occurs at definite rate and always follows exponential law. Thus, the number of atoms disintegrating at any time is proportional to the number of atoms of the isotope present at that time. The exponential curve of radioactive decay is expressed mathematically as:

$$\frac{-dN}{dt} = \lambda N$$

where the rate of change of the number of atoms is proportional to the number of atoms (N) present multiplied by the decay constant (λ) (Fig. 6.1).

Table 6.1: Half life of different Isotopes

Isotope	*Half life (t1/2)*	*Emission*
^{14}C	5568 years	β-rays
^{3}H	12 years	β-rays
^{35}S	87 days	β-rays
^{59}Fe	45 days	β and γ-rays
^{32}P	14.2 days	β-rays
^{24}Na	15 hours	β and γ-rays

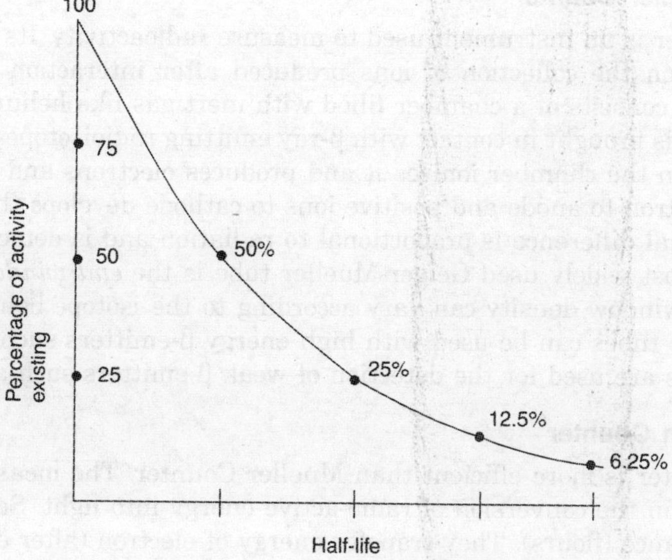

Fig. 6.1 Exponential curve of radioactive decay

6.2.5 Radioactive Units

The basic unit of radioactive decay is represented by curie (Ci). It can be defined as the amount of isotope that undergoes 3.7 × 10^{10} disintegration per second. Most commonly used decay units are millicurie (MCi) and microcurie (μci).

Another unit of radiation is Roentgen. It is a unit of radiation exposure and is based on ionisation produced by radiation. It can be defined as the amount of radiation that produces ions carrying one electrostatic unit of electricity.

6.3 PRODUCTION OF RADIOISOTOPES

The isotopic forms of an element exist in nature. They can be extracted and purified from their natural sources. Due to several constraints, isotopes are produced in nuclear reactors. Neutron (n) produced in nuclear reactors combine with an isotope of an element to produce a radio isotope. For example, ^{14}C is produced from ^{14}N when it combines with a neutron

$$^{14}\text{N} + {}^{1}\text{N} \longrightarrow {}^{14}\text{C} + \beta\text{ - rays}$$

In another method, isotopes are produced in a cyclotron by bombardment of the nucleus of an element with proton.

6.4 MEASUREMENT OF RADIOACTIVITY

Radioactivity can be measured with the help of device known as counters. There are two types of counters employed for radioactivity measurement. These are extensively used for measurement of radioisotopes that emit β-rays.

6.4.1 Geiger Mueller Counter

The Geiger counter is an instrument used to measure radioactivity. Its working principle is mainly based on the collection of ions produced after interaction of radiation with matter. It mainly consists of a chamber filled with inert gas like helium or argon. When gas filled counter is brought in contact with β-ray emitting radioisotopes, radiation passes through the gas in the chamber ionizes it and produces electrons and positive ions. The movement of electron to anode and positive ions to cathode develops the potential difference. This potential difference is proportional to radiation and is detected as count.

One of the most widely used Geiger-Mueller tube is the *end window tube* (Fig. 6.2). This count tube window density can vary according to the isotope being detected. Thus, thick end window tubes can be used with high energy β-emitters such as ^{32}P while thin end window tubes are used for the detection of weak β-emitters such as ^{14}C.

6.4.2 Scintillation Counter

Scintillation counter is more efficient than Mueller Counter. The measurement or radioactivity is based on the conversion of radio-active energy into light. Scintillators are the fluorescent substance (flours). They transfer energy of electron (after excitation) when it is hit by radiation. The light emitted by scintillation can be detected by the use of a photomultiplier which converts photons into electric pulse. Some of the common scintillators employed in scintillation counter are sodium iodide crystals (γ-emitter), zinc sulphide (silver activated, α-emitter) anthracene (β-emitter) and stilbene. When scintillators are hit by ionising particles, they absorb energy and undergo excitation.

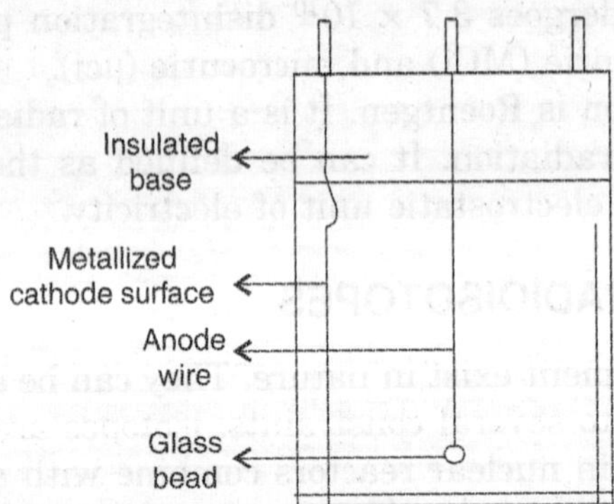

Fig. 6.2 Geiger-Mueller tube

Later, energy is emitted as a pulse of visible or ultraviolet light. The emitted light is then observed by a photomultiplier tube.

6.4.2.1 Types of Scintillation Counter

There are two types of scintillation counter — (1) Solid scintillation counter (2) Liquid scintillation counter.

6.4.2.1.1 Solid scintillation counter: It is otherwise known as external scintillation counter. In this type, the sample is placed adjacent to a scintillator (crystal of fluorescent material). The crystals may be sodium iodide, zinc sulphide and b anthracene depending upon the type of radioactive emission. These crystals are placed near to a photomultiplier connected to a high voltage supplier and also a scalar. Solid scintillation is useful for γ-emitting isotopes (Fig. 6.3).

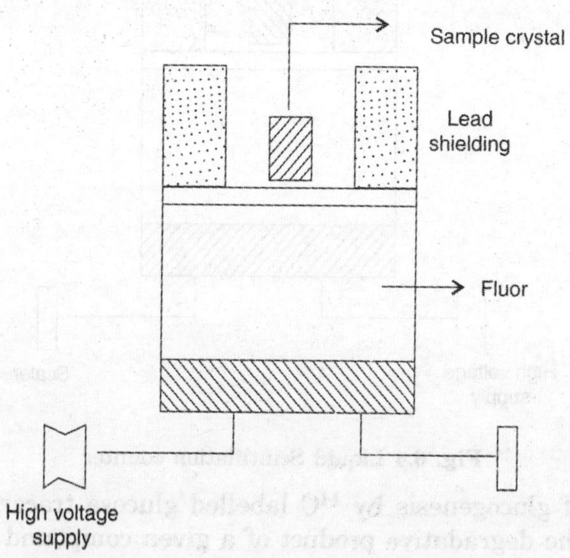

Fig. 6.3 Solid Scintillation Counter

6.4.2.1.2 Liquid scintillation counter: It is otherwise known as internal scintillation counter. The sample is mixed with a solvent to the anthracene which contains the appropriate scintillation in solution. The liquid scintillation system is useful for β-emitting isotopes like ^{3}H, ^{14}C and ^{35}S. The isotopes are predominantly used in biological system.

In liquid scintillation system, when radioactivity is made to bombard with organic solvent, they fluorescence and emit a light of very short wavelength. This cannot be detected by phototubes. However, if a compound is dissolved in the solvent which can accept the energy from the solvent and fluoresce at a longer wave length the emitted light is then efficiently detected. Such compound is called primary flour like 2.5 diphenyl-oxazole (Fig. 6.4).

6.5 APPLICATIONS OF RADIOISOTOPES

Radioisotopes have been using in various biological field and are highly diversified such as agriculture, tumour therapy, immuno diagnostic and food industry.

6.5.1 Isotopes in Tracer Technique

In order to understand biochemical reactions it is very essential to use radioisotopes in tracking metabolic pathways. For example, certain compounds like glucose are labelled with ^{14}C (^{14}C glucose) and aminoacids (^{15}N-glutamate) using isotopes as tracers. Several major metabolic pathways have been elucidated like carbon dioxide fixation in plants during photosynthesis, central metabolic pathways, synthesis of nucleic acid, process of DNA replication. In medical biochemistry, these are widely used in studying the action of

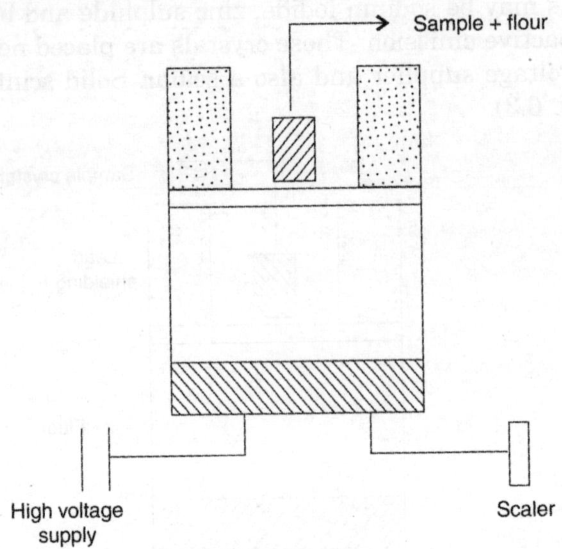

Fig. 6.4 Liquid Scintillation counter

branching enzymes of glucogenesis by ^{14}C labelled glucose tracer. An isotopic tracer is used to understand the degradative product of a given compound like uric acid, an end product of purine degradation has been established using ^{14}C-labelled guanine as the tracer. The rate of synthesis of DNA or replication has been determined using an ^{3}H labelled thymidine tracer. It can throw light on the metabolic abnormalities and sites of destruction. It also provides information on the rate of absorption and defects of a given mineral using labelled mineral (^{59}Fe) in intestine.

Isotopes are used in the measurement of body mass, body water and extracellular fluid using ^{3}H labelled tritium oxide, ^{14}C labelled uric acid and ^{51}Cr for erythrocyte volume.

6.5.2 Isotopes in Radioimmuno Assay (RIA)

It is extremely difficult to measure the presence of hormones in the blood as their concentration are in nano or picomole. Applying isotopes in a technique known as radioimmuno assay (RIA) it is possible to detect picomoles or femto mole amounts of hormones or any other compounds. In RIA, antigens are labelled in antigen-antibody complex. The hormone is allowed to react with labelled antigen-antibody complex. The amount of hormone in the sample is calculated by measuring the amount of labelled antigen displaced.

6.5.3 Isotopes in Organ Functions and Imaging

Radioisotopes are used to study functions of thyroid and kidney. Using ^{131}I the thyroid function is assessed.

Rate of synthesis and elimination of thyroxine hormone can be measured using radiolabelled hippuric acid, kidney function is assessed.

In organ imaging, isotopes such as ^{131}I is used to obtain images of the lung and thyroid.

6.5.4 Radiation and Isotopes in Agriculture

Radioisotopes are indispensable in studying the mode of ion uptake by root system from the soil. Radiation is employed to induce mutations in order to improvise crop productivity in agriculture. Calibrated dosage of γ-irradiated seeds or plants are employed in process of mutation breeding.

6.5.5 Isotopes in Food Industry

In food industry, isotopes are used to increase the shelf life of foods and dairy products. For example, pasteurised milk exposed to radiation has a longer shelf life. Several foods are sterilised by exposing to radiation before packaging is done. Radiosterilised foods have an increased shelf life.

6.5.6 Isotopes in Cancer Therapy

Radiation has been used in the treatment of cancer since long time. Tumour cells are sensitive to radiation compared to normal cells. Isotopes such as ^{131}I is used in the treatment of thyroid cancer and ^{60}CO is extensively used in the treatment of tumours located deep inside the body. For skin cancer and leukemia ^{32}P is widely used.

6.5.7 Autoradiography and probes

Autoradiography is widely used in cell biology and molecular biology. It is based on the interaction of radiation from the radio isotopes with photographic emulsion. The sample containing the radio labelled compound is brcught closer to the photographic film. Ionising radiation from the sample falls on the photographic film and interacts with photographic emulsion. The dark area on the film indicates the localisation of radio isotope in the sample.

QUESTIONS

1. What are radio isotopes? Explain their use in tracer technology. (BU Oct 1998)
2. Discuss the role of radio isotopes in biological system.
3. What is scintillation counter? Explain the types and their significance.

SHORT QUESTIONS

1. Radioactive decay
2. Geiger Mueller counter
3. Scintillation Counter
4. Solid Scintillation Counter
5. Liquid Scintillation Counter